COASTAL FISH
Identification
CALIFORNIA to ALASKA

PAUL HUMANN
With Photographers
HOWARD HALL & NEIL McDANIEL

EDITED BY
NED DELOACH

NEW WORLD PUBLICATIONS, INC.
Printed by
Paramount Miller Graphics, Inc.
Jacksonville, Florida U.S.A.

PHOTO CREDITS

Many underwater photographers contributed their work to this collection. I appreciate their efforts and assistance in making this book as comprehensive as possible. They include: **Stephen Benavides**, 100m, 101m, 153t; **Wayne Brown** 97m; **Jay Carroll**, 83b, 97b; **Dr. Mark Chamberlin**, 65t, 68mr, 69t, 99m, 117t, 119b, 123b, 125t; 128b **Gary Cissne**, 114ml&r, 115b; **Brandon Cole**, 82mr, 85m, 91m, 93t&m, 131b, 141t&m; **Mark Conlin**, 16, 17, 19, 22m, 23t&b, 25b, 27t, 35b, 37m, 41m, 57m, 63b, 64m, 65m, 69b, 71b, 82ml, 83t&m, 89m, 98mr, 99t&b, 103m, 105t&m, 123m, 129t&b, 130b, 131t, 139t&m, 143b, 149b, 155t, 165tm&b, 169b, 173b, 177b, 179t&b, 181t&b, 185t, 187m, 193m, 199m, 203m&b, 205m; **Ned DeLoach**, 177t; **Dr. David Hall**, 81t; **Howard Hall**, 18tr&b, 27b, 29t&b, 30m, 31m, 32b, 37t&b, 39m, 41b, 45t&b, 47m, 49t, 52m, 53m, 56mr, 68ml, 73b, 79t&m, 97t, 98ml, 102b, 103t, 108b, 115t, 123t, 145t, 149t, 151tm&b, 152m, 155m&b, 157t, 169t&m, 171b, 175m, 179m, 181t, 185b, 187t, 191tm&b, 194m, 195m&b, 197m, 199b, 201b, 205t; **Michele Hall**, 122b; **Rick Harbo**, 109t; **Richard Herman**, 40b, 183b; **Neil McDaniel**, 18tl, 27m, 31b, 33tm&b, 35t&m, 39b, 41t, 43tm&b, 47t, 49m, 55m&b, 57t&b, 58m, 59m&b, 61b, 62m, 65b, 66m, 67tm&b, 69m, 73t&m, 75t&m, 81m&b, 85t&b, 86m, 87tm&b, 89t&b, 90m, 91t, 93b, 104m, 105b, 106b, 111t&m, 113m, 107tm&b, 109b, 117m, 119t, 125m&b, 127tm&b, 131m, 135tm&b, 137t&m, 139b, 141b, 142m, 143t&m, 149m, 157m, 178b, 183t&m, 193t&b, 197t, 201t&m; **John Pennington** 31t, 71m, 80m; **Jeffery Rotman**, 23m, 25t; **Marty Snyderman**, 45m, 79b, 115m, 129m; **Graeme Teague**, 63m; **Jill Wallin**, 101b; **Karl Wallin**, 96b. The remaining pictures were taken by the author, **Paul Humann**, cover, 1, 10, 11, 25m, 29m, 39t, 47b, 53t&b, 55t, 56ml, 59t, 60m, 61t&m, 63t, 70m, 71t, 75b, 91b, 101t, 103b, 107t, 109t, 113t&m, 117t, 119m, 137b, 145m, 150m 153m&b, 154b, 156m, 157b, 159tm&b, 161t&m, 167tm&b, 171t&m, 173t&m, 175t&b, 177m, 185m, 195t, 197b, 199t, 203t, 205b.

CREDITS

Editor: Ned DeLoach
Copy Editors: Anna DeLoach & Mary DeLoach
Layout & Design: Paul Humann & Ned DeLoach
Art Director: Michael O'Connell
Drawings: Michael O'Connell & Michael Elliott
Color Separations: Buddy Waggoner & Kit Keating of Paramount Miller Graphics, Jacksonville FL
Printed by: Paramount Miller Graphics, Jacksonville, FL
First Printing: 1996

PUBLISHER'S CATALOGING IN PUBLICATION DATA

Humann, Paul
Costal Fish Identification: California to Alaska/by Paul Humann; with photographers Howard Hall and Neil McDaniel.
p. cm.
Includes index.
ISBN: 1-878348-12-4 (comb binding)
ISBN: 1-878348-17-5 (library hardcover)

1. Marine fishes — Pacific Coast (U.S.)—Identification. 2. Marine fishes—Pacific Coast (B.C.)—Identification. 3. Fishes—Pacific Coast (U.S.)—Identification. 4. Fishes—Pacific Coast (B.C.)—Identification. I. Hall, Howard, 1949- II. McDaniel, Neil, III. Title.

QL623.4.H86 1996   597.092'1643   QBI96-20303   LCN-96-068131

Copyright © 1996 by Paul Humann
All right reserved.
No part of this work may be reproduced or transmitted in any form or by any means, electronic or mechanical, including photocopying and recording, or by any information storage or retrieval system, except as may be expressly permitted by the 1976 Copyright Act or in writing from the publisher.

Publisher: New World Publications, Inc., 1861 Cornell Road, Jacksonville, FL 33207, (904)737-6558

Personal Acknowledgments

The concept and design of this series of marine life identification books were the result of considerable encouragement, help, and advice from many fiends and acquaintances. I wish to express my sincere gratitude to everyone involved. Naturally, the names of a few who played especially significant roles over the years come to mind. They include Patricia and Richard Collins, John & Marion Bacon, Mike Bacon, Guy Beard, Ken Marks, Anna DeLoach and Mary DeLoach. Feodor Pitcairn dragged me along (I was,and still am, a warm water sissy.) on several dive excursions to Canada that produced many of the pictures for this book and opened my eyes to the wondrous beauty of the Pacific Northwest. Jim Borrman and Bill MacKay taught me how to use a dry suit without killing myself. Of course, I must mention my partner, collaborator, editor and best friend Ned DeLoach. Without his encouragement and most able assistance none of these identification books would have been published. Finally, I must include the young man that "really" runs our publishing business, Eric Riesch. He is involved in nearly all our decisions. Without his input and assistance our business would probably collapse!

Obviously, this book would have never been published without the cooperation of photographers and good friends, Howard & Michele Hall and Neil McDaniel. Their extensive collection of superb fish portraits from the North American Pacific Coast made the project possible. My most sincere thanks to them for opening their valuable files for my use.

Although I have dived the Canadian Pacific Northwest and California waters on numerous occasions, my firsthand knowledge of many species was far from complete. Howard Hall, Michele Hall, Neil McDaniel, Mark Conlin, Marty Snyderman and Brandon Cole freely shared their considerable knowledge of the habitats, behaviors and reactions of divers of many species. Michele Hall also graciously put me in touch with several photographers that had pictures of species absent from Howard's extensive collection.

Scientific Acknowledgments

Special recognition must be given to the ichthyologists who gave freely of their time, advice and knowledge in confirming or providing identifications and supplemental information. Without their most generous assistance, the value of this book would be greatly diminished and the number of included species reduced. Every attempt has been made to keep the text and identifications as accurate as possible; however, I'm sure a few errors crept in, and they are my sole responsibility.

John McCosker, Ph.D., California Academy of Sciences, San Francisco, CA
Our friendship goes back to 1977 when we dived, with reckless abandon, well below safe diving limits, at night without lights, to capture some of the first living specimens of the flashlight fish, *Kryptophanaron alfredi* — an event we later pegged the "Krypo Caper." John's enthusiasm for studying fish is boundless and infectious. To this day we continue to enjoy diving and researching together. John was my primary scientific reference for this book. He is the former Director of The Steinhart Aquarium, who today happily pursues his passion for research on a full time basis.

Richard Rosenblatt, Ph.D., Scripps Institution Of Oceanography, La Jolla, CA
Dick was also a member of the Krypo Caper expedition. Recently we enjoyed another research excursion to Galapagos. I never cease to be amazed at Dick's knowledge of fishes; he constantly pulls up facts and figures that one would expect only to be stored in a computer. His skill identifying species from photographs alone is a great asset. He reviewed the sculpin photographs for accuracy and helped sort out several other difficult-to-identify species.

M. James Allen, Ph.D., Southern California Coastal Water Research Project, Westminster, CA
I have never met Dr. Allen, but his help was essential to this book. Many flatfishes appear quite similar and are often extremely difficult to identify to species. In checking with other ichthyologists who might assist is sorting out this group the answer always came back, "If anyone would know, it would be Jim Allen." Dr. Allen not only reviewed the pictures, but also included copious notes about the visual clues he used in making the identifications. These became the basis of the DISTINCTIVE FEATURES I used for the flatfishes. He also reviewed photographs of several additional problem species.

Also assisting:
William Smith-Vaniz, Ph.D., National Fish & Wildlife Service, Gainesville, FL
Douglas Long, Ph.D., California Academy of Sciences, San Francisco, CA
Jack Engle, Ph.D., Marine Science Institute, University of California, Santa Barbara, CA

About The Author & Photographers

    Paul Humann took his first underwater photographs in the early 1960's. By the late 70's several of his fish portraits were published in *Skin Diver's* memorable *Fish of the Month* series. His hobby became a way of life in 1970 when he left a successful law practice in Wichita, Kansas, to become captain/owner of the *Cayman Diver*, the Caribbean's first live-aboard dive cruiser. His next eight years were spent documenting the biological diversity of the Caribbean's coral reefs. His dive charter business was sold in 1979 to gain the freedom to study and photograph marine creatures around the world.

    His desire to learn and teach about sea life has been the catalyst for numerous magazine articles, four large-format photographic books and the award-winning *Reef Set* — a comprehensive, three volume visual identification guide to the marine life of Florida, the Bahamas and Caribbean. The now famous marine life trilogy includes *Reef Fish, Reef Creature* and *Reef Coral Identification*. A fourth visual marine life identification guide, *Reef Fish Identification — Galapagos,* was published in 1993 and *Snorkeling Guide to Marine Life* appeared in 1995.

    This pioneering work, establishing visual identification criteria for marine creatures, requires much more than the difficult task of capturing each species on film. Long hours of observation, documentation, cataloguing and corresponding with dozens of marine taxonomists are each essential steps in the long process. Thanks to his efforts, it is now possible for underwater naturalists to make valid, non-impact biodiversity assessments of reef ecosystems. His concern for the welfare of sea life has also led to the founding of the Reef Environmental Education Foundation (REEF) — an organization of recreational divers that regularly monitor marine fish populations. Paul resides in south Florida when he is not traveling the world with cameras and dive gear in tow.

Howard Hall, a native of southern California, began diving as a teenager. He successfully combined his fascination for the marine environment with the necessity to earn money for college by teaching diving lessons. After graduating with a degree in zoology, he concentrated his interest on underwater photography. It was the right decision; he quickly established himself as one of the world's best known marine wildlife photographers. His superb photographs and entertaining narratives of his many exciting encounters with marine creatures appear in dozens of publications at home and abroad. In 1982 he authored *Successful Underwater Photography*, one of the first and most widely used marine photography reference books ever published.

In recent years his career has evolved toward marine life cinematography. His impressive production credits include films for the PBS series *Nature* and *National Geographic Specials*. Recently he directed the first underwater IMAX 3D natural history film. His outstanding underwater film work has garnered six television Emmys and numerous natural history film honors.

Film production provides the means for Howard to fulfill his dream of diving, exploring and working in remote corners of the earth. Even though most of his present underwater time is spent with bulky film systems, his still camera is never far from reach.

Neil McDaniel received a degree in Marine Zoology from the University of British Columbia in 1971. While attending college he learned to dive and immediately became captivated with the natural history of marine life of the Northwest Pacific Coast. It wasn't long before he discovered that underwater photography was a necessary tool to document the behavior of marine creatures, many of which were seldom, if ever, seen except by divers.

His career includes eight years as a biological technician for Fisheries and Oceans Canada and another seven years as editor of Canada's *Diver Magazine*. He contributes natural history photographs and articles to many international magazines and books and is currently preparing a pictorial guide to the marine invertebrates of the West Coast. His travels enable him to dive extensively in tropical regions of the world, but he still finds the cold, green waters of the Pacific Northwest among the most intriguing underwater realms.

Currently based in Vancouver, Canada, Neil continues to pursue his varied interests in photojournalism, underwater cinematography, video production and biological consulting.

Contents

How To Use This Book

    Identification Groups 10
    Names 10
    Size 11
    Depth 11
    Distinctive Features 11
    Description 12
    Variations & Phases 13
    Abundance & Distribution 14
    Map – *Alaska & British Columbia* 14
    Map – *Washington, Oregon & California* 15
    Habitat & Behavior 17
    Reaction to Divers 19
    Similar Species 19
    Note 19

Nine Identification Groups

1. Heavy Body/Large Lips   20-49

Sea Bass

Giant Sea Bass

Rockfish

2. Bulbous, Spiny-Headed Bottom Dwellers   50-75

Scorpionfish

Sculpin

3. Eels & Eel-Like Bottom-Dwellers   76-93

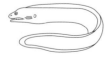
Moray

Conger Eel

Cusk-eel

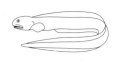
Wolf-eel

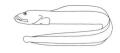
Wrymouth

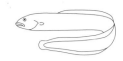
Prickleback

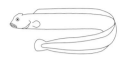
Cockscomb

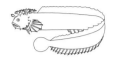
Warbonnet

Eelpout

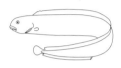
Gunnel

Hagfish

Quillfish

4. Elongated Bottom-Dwellers   94-119

Fringehead

Pikeblenny

Kelpfish

Greenling

Cod

Lizardfish

Tilefish

Brotula

Combtooth Blenny

Ronquil

5. Flatfish/Bottom Dwellers   120-131

Turbot

Sole

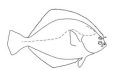
Flounder

Halibut

Sandab

6. Odd-Shaped Bottom Dwellers   132-145

Poacher

Bigeye

Frogfish

Snailfish

Clingfish

Midshipman

Pipefish & Seahorse

7. Odd-Shaped & Other Swimmers   146-161

Mola

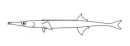
Stickleback

Cardinalfish

Wrasse

Butterflyfish

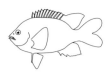
Damselfish

Ratfish

Triggerfish

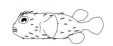
Puffer

8. Silvery Swimmers 162-187

Herring

Anchovy

Silverside

Jack

Mackerel

Billfish

Dolphin

Barracuda

Machete

Flyingfish

Bonefish

Goatfish

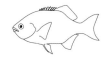
Surfperch

Grunt

Sea Chub

Croaker

9. Sharks & Rays 188-205

Shark

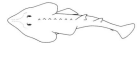
Guitarfish

Ray

How To Use This Book

Giant Kelp Forest

Identification Groups

Trying to identify a specific fish from the more than 300 species that inhabit the coastal waters of the North American Pacific Coast can be a perplexing task. To help simplify the process, families are arranged into nine color-coded and numbered ID Groups. Each group is distinguished by similar physical and/or behavioral characteristics that can be recognized underwater. Although there are a few anomalies, most species integrate easily into this system.

The ID Groups and their families are displayed on the Contents pages. Each group's similar physical characteristics are listed at the beginning of its chapter. It is important for beginning fishwatchers to become familiar with the make-up of each ID Group so that they can go quickly to the correct chapter to start the identification process.

The next step is to learn to recognize the major families that comprise each ID Group. Families are scientific groupings based on evolutionary sequence, and, consequently, typically have similar physical characteristics. An overview of the major family's behavioral and physical characteristics (that are observable by divers) is presented at the beginning of each chapter. The total number of species included, along with profile drawings of representative body shapes, is also given.

Names

Information about each fish begins with the common name (that used by the general public). In the past, using common names for identification was impractical because the same fish species was often known by different common names in different areas. For example, Yelloweye Rockfish, Turkey-red Rockfish, and Rasphead Rockfish all

all refer to the same fish. In 1948, The American Fisheries Society helped standardize common names by publishing a preferred list that is updated every ten years. Their recommendations are used in this book. Common names are capitalized to help note their position in the text, although this practice is not considered grammatically correct.

Below the common name, in italics, is the two-part scientific name. The first word (always capitalized) is the genus. The genus name is given to a group of species which share a common ancestor, and usually have similar anatomical and physiological characteristics. (For example, in the Scorpionfish family all the rockfishes are in the genus *Sebastes*, and look and behave much the same). The second word (never capitalized) is the species. A species includes only animals that are sexually compatible and produce fertile offspring. (Continuing our example of rockfishes, the Yelloweye Rockfish is *Sebastes ruberrimus*, while the China Rockfish is *Sebastes nebulosus*, and so on.) Most species have one or more visually distinct features that separate them from all others. Genus and species names, rooted in Latin and Greek, are used by scientists throughout the world.

Common and scientific family names are listed next. Occasionally a group of species within a family has acquired a name more familiar to the public than their common family name. An example would be rockfish which are members of the scorpionfish family. When this occurs, the group name is substituted for the common family name in the Quick-Reference Index, Content pages, and at the top of identification text. However, the correct common family name is always listed next to the species photograph.

Size

The general size range of the fish that divers are most likely to observe, followed by the maximum recorded size.

Depth

In most cases , the depths included for a particular species are those reported in scientific literature. Occasionally, species are sighted outside this range. Personal observations by contributing photographers have also been included where appropriate.

Bull Kelp Forest

Distinctive Features

Colors, markings and anatomical differences that distinguish the fish from similar appearing species. In most cases, these features are readily apparent to divers, but occasionally they are quite subtle. When practical, the locations of these features are indicated by numbered arrows on the profile drawing next to the photo-

Description

A general description of colors, markings and anatomical features. The information included in DISTINCTIVE FEATURES is not repeated in this section unless it is qualified or expanded upon.

Colors — The colors of many species vary considerably from individual to individual. In such situations, the DESCRIPTION might read: "Vary from reddish brown to olive-brown or gray." This means that the fish could be any of the colors and shades between. Many fish also have an ability to pale, darken, and change color. Because of this, color alone is rarely used for identification. An exception is the Garibaldi, pg. 154-55, whose brilliant orange color is distinctive.

Markings — The terminology used to describe fish markings is defined in the following drawings.

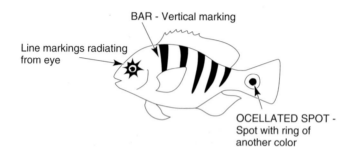

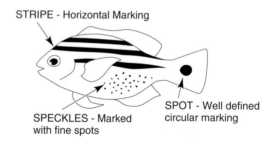

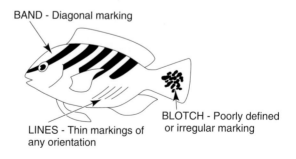

Anatomy — Anatomical features are often referred to as part of the identification process. The features used in this text are pinpointed in the following drawings.

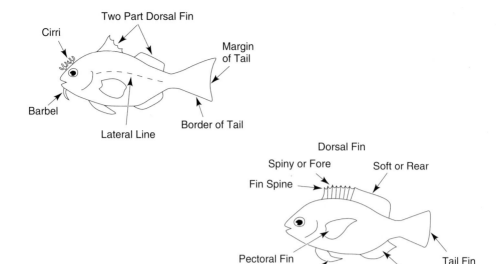

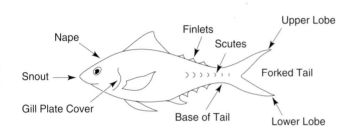

Variations & Phases

    Many species are exhibited in more than one photograph. This is necessary to show the variations in color, markings and shape that occur within the species.

    Occasionally, the maturation phases of certain fish are so dramatic that they may be confused as a different species altogether. These phases can include JUVENILE, INTERMEDIATE and ADULT. The marked difference between the juvenile and adult Yelloweye Rockfish are displayed on pg. 30 - 31. Wrasse are somewhat different; they have a juvenile phase (JP), adults, known as the initial phase (IP), and a terminal phase (TP), which includes only sexually mature males. Terminal males are the largest and most colorful phase. The California Sheephead, pg. 150-151, illustrate all three phases. When the maturation phase is not given, the photograph is of an adult. No attempt has been make to include juveniles that closely resemble adults, or that live in habitats not frequented by divers.

    Many fishes are able to alter both color and markings, either for camouflage or with mood changes. If these variations confuse identification, they may also be pictured. Note the Buffalo Sculpin, pg. 58 - 59, and the Giant Kelpfish, pg. 96 - 97. In a few species, the appearance between the male and female also differs, such as the Painted Greenling, pg. 100 - 101, and the Dolphin, on pg. 172 - 173.

Abundance & Distribution

Abundance refers to a diver's likelihood of observing a species in its normal habitat and depth range on any given dive. Because of reclusive habits and other factors, this does not always present an accurate portrait of actual populations. Definitions are as follows:

Abundant — At least several sightings can be expected on nearly every dive.
Common — Sightings are frequent, but not necessarily expected on every dive.
Occasional — Sightings are not unusual, but are not expected on a regular basis.
Uncommon — Sightings are unusual.
Rare — Sightings are exceptional.

Distribution describes where the species is found geographically. Its distribution along the North American Pacific Coast is given first. If the abundance within this range varies, the locations are listed in sequence from areas of most sightings to those of least sightings. Concluding this section is additional distribution of the species beyond the range of this book; however, abundance information is not included for these areas.

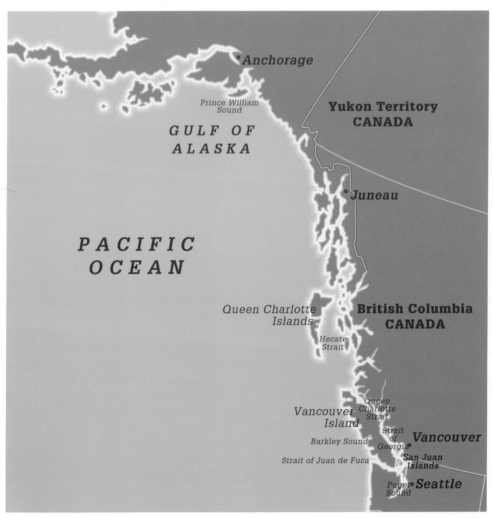

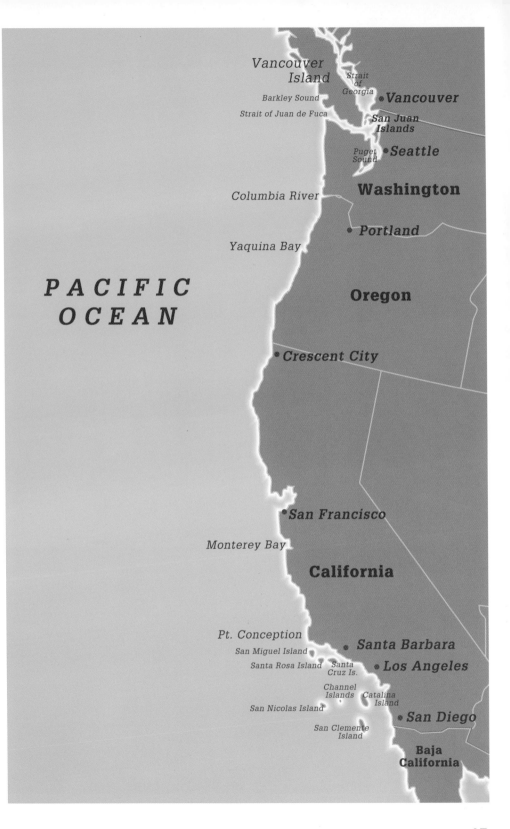

Aggregation

Habitat & Behavior

Habitat is the type of underwater terrain where a particular species is generally found. Habitats frequented by divers, such as kelp beds, adjacent areas of sand and rubble, sea grass beds and the vicinity of docks and pilings, have been emphasized.

Behavior is the fish's normal activities that may be observed by a diver that helps in identification. The schooling behavior of some species is often an important identification clue. Types of schooling are detailed below.

Schooling — Many fish congregate in groups, commonly called schools. The two primary reasons for this behavior are predator protection and cooperative hunting. It is theorized that hunted species pack together and move in unison so that predators have a difficult time picking out and attacking a single target. In the confusion of numbers all members of the school may survive the attack. If, however, a fish becomes separated from the school, the predator can concentrate on the lone fish and the attack is often successful. Some predators also school. They facilitate hunting by cooperating with each other in separating and confusing potential victims.

There are several types of fish congregations. In polarized schools, all fish swim together in the same direction, and at the same speed, at an equal distance apart. Those forming non-polarized schools, or simply "schools," stay close together, but do not show the rigid uniformity displayed by polarized schools. Fish often come together for reasons other than protection or hunting. These loose gatherings, often including a mixture of species, are referred to as aggregations. Aggregations may drift together in a shaded opening in kelp forests, or can be attracted to an areas with a plentiful source of food.

Polarized School

School/Aggregation

School

Aggregation

Non-polarized School

Reaction To Divers
This information relates to the fish's normal reaction when confronted by a diver, and what a diver can do to get a closer look.

Similar Species
Occasionally there are similar appearing species not pictured. These fish, in most cases, are rarely sighted by divers. Information is given to help distinguish these fish from the pictured species.

Note
Occasionally supplementary information helpful in the identification process or of general interest is included as a note. Recent changes in a species classification and additional (but not preferred) common names also appear in the section when appropriate.

IDENTIFICATION GROUP 1

Heavy Body/Large Lips
Sea Bass – Rockfish

This ID Group consists of fish with stout, well-built, "bass-like" bodies. They have large mouths and lips, and usually a jutting lower jaw. The long, continuous dorsal fin is noticeably divided into two parts. The foredorsal is supported by spines that can be held erect or lowered; flexible rays form the structure for the rear dorsal.

FAMILY: Sea Bass — Serranidae
Temperate Bass — Percichthyidae
7 Species Included

Sea Bass
(typical shape)

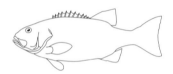

Giant Sea Bass

Bass and groupers are the best known members of the sea bass family. All members of this large family, consisting of over 375 species worldwide, have strong, stout bodies, large mouths and lips, and, in most cases, projecting lower jaws. The majority, including the smaller basslets and soapfish, inhabit tropical waters. Along the Pacific coast of North America sea basses range in size from the Gulf Grouper that can grow over six feet to the Spotted Sand Bass that is less than two feet in length. Larger species are generally solitary carnivores that live near the bottom where they lurk in the shadows of reefs, ledges, wrecks, kelp and other marine growth.

Although cumbersome in appearance, these stocky fish can cover short distances with sudden bursts of speed. A powerful suction created when they open their cavernous mouths instantly pulls in fish, crustaceans and other prey. Food is held securely by thousands of small, rasp-like teeth that cover the jaws, tongue and palate. The prey is swallowed whole. Many groupers are hermaphroditic, beginning life as females but changing to males with maturity.

Many sea bass are difficult to identify to species because of their ability to lighten, darken and change colors and even markings. However, each species displays distinctive markings or physical characteristics that, with careful observation, allow underwater identification.

The Giant Sea Bass from the temperate bass family Percichthyidae is also included in ID Group 1 because of its visual similarity to sea bass. This great fish can grow to over seven feet in legnth.

FAMILY: Rockfish — Scorpaenidae
30 Species Included

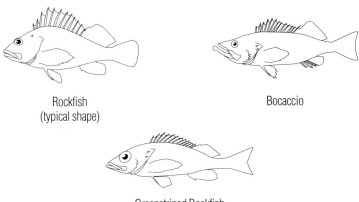

Rockfish
(typical shape)

Bocaccio

Greenstriped Rockfish

    Although rockfish, from the genus *Sebastes*, are members of the scorpionfish family, they are included here because in appearance, size, and behavior they more closely resemble sea bass. *Sebastes*, which means "magnificent," is appropriate for the beautiful colors and markings of many rockfishes. About 60 members of the genus inhabit the waters of the North American Pacific Coast. Their bottom dwelling relatives, scorpionfish, are discussed in the next ID Group — Bulbous, Spiny-Headed Bottom-Dwellers.

    Rockfish are bass-like with heavy, more or less compressed bodies, large mouths, jutting lower jaws and prominent lips. Their size varies from around six inches to slightly over three feet. Most have spines on their gill plate covers and heads. Like other scorpionfish, the foredorsal, anal and ventral fin spines of rockfish are venomous, but only slightly, compared to the stonefish and lionfish of the tropical Indo-Pacific.

    Rockfish swim or drift above the bottom in several habitats, including reefs, rocky recesses, kelp forests and other marine growth. They occasionally rest on the bottom propped up by their fins. A few species form aggregations, or schools, in open water. All rockfishes are ovoviviparous (eggs, which are fertilized and develop internally, are ejected just before hatching).

    Many rockfish are easily identified by their bright colors and bold markings. Others, however, are rather drab with only subtle differences, which makes them difficult to distinguish from similar appearing genus members.

Temperate Bass – Sea Bass

DISTINCTIVE FEATURES: Large size. **1. Sizable black spots over dark brown to gray (except very large individuals that tend to have uniform coloration). 2. Low profile foredorsal fin; soft dorsal noticeably taller.**
DESCRIPTION: Large mouth and bulky body. Juveniles red undercolor, changing to shades of brown, then dark brown or gray as they mature.
ABUNDANCE & DISTRIBUTION: Rare southern California; very rare north to northern California. Also south to Gulf of California. Formerly more common, but numbers reduced to near extinction by overfishing; possession unlawful in California.
HABITAT & BEHAVIOR: Inhabit rocky bottoms, in vicinity of rocky outcroppings and in kelp forests. Drift in shaded areas blending with surroundings. Larger individuals tend to be in deeper water.
REACTION TO DIVERS: Appear unafraid and often curious; usually allow close view when approached with slow nonthreatening movements. May bolt into cave or recess and return to entrance to peer out.
NOTE: Also commonly known as "Jewfish," "Black Bass," or "Black Sea Bass."

Giant Sea Bass
Large individual displaying uniform gray color.

DISTINCTIVE FEATURES: 1. Margin of tail deeply serrated and ragged. (Similar Gulf Grouper [next] distinguished by smooth tail margin.) **2. Dark lines radiate from eye (not always obvious when displaying dark pattern).**
DESCRIPTION: Shades of brown to gray. Color frequently solid, but can display scrawls, blotches or oval markings on back. Anal fin rounded.
ABUNDANCE & DISTRIBUTION: Rare central to southern California. Also south to Peru. Threatened by overfishing; possession unlawful in California waters.
HABITAT & BEHAVIOR: Wide range of habitats; often in or near shallow estuaries.
REACTION TO DIVERS: Appear unafraid; usually allow a slow nonthreatening approach to about 15 feet before bolting.

Heavy Body/Large Lips

GIANT SEA BASS
Stereolepis gigas
FAMILY:
Temperate Bass –
Percichthyidae

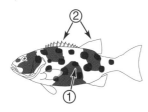

SIZE: 2-4., max. 7 1/2 ft.
DEPTH: 20-150 ft.

Giant Sea Bass
Dark, nearly solid color pattern.

BROOMTAIL GROUPER
Mycteroperca xenarcha
FAMILY:
Sea Bass – Serranidae

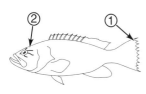

SIZE: 1 1/2 - 2 1/2 ft.,
max. 4 ft.
DEPTH: 0-70 ft.

Sea Bass

DISTINCTIVE FEATURES: 1. Straight vertical outer edge on anal fin. (Similar Broomtail Grouper [previous] distinguished by rounded anal fin and ragged tail margin.) **2. Dark lines radiate from eye (not always obvious when displaying dark pattern).**
DESCRIPTION: Shades of brown to gray with dark blotches on back (not always obvious when displaying dark pattern); generally pale undercolor boldly displays markings, but can rapidly darken to nearly a solid color pattern; fins often have light edge. Smooth margin of tail square-cut or slightly indented.
ABUNDANCE & DISTRIBUTION: Rare southern California. Also south to Baja and Gulf of California. Threatened by overfishing; possession unlawful in California waters.
HABITAT & BEHAVIOR: Inhabit rocky reefs and outcroppings; often drift inside caves and under overhangs.
REACTION TO DIVERS: Appear unafraid; usually allow a slow nonthreatening approach to about 15 feet before bolting.

DISTINCTIVE FEATURES: 1. Large, pale to white, oval to rectangular spots or blotches on back. 2. First two spines of foredorsal fin short; spines three to five are of nearly equal length. (Similar Spotted Sand Bass [next] and Barred Sand Bass [following], third spine is distinctly longer.)
DESCRIPTION: Back and sides mottled in solid shades of brown to black; belly pale. Tail square-cut.
ABUNDANCE & DISTRIBUTION: Common to occasional southern California; uncommon to rare north to Washington. Also south to southern Baja.
HABITAT & BEHAVIOR: Young and small to medium-sized adults inhabit kelp beds, rocky inshore areas and seaweed flats; large adults inhabit deeper patch reefs and areas of sand. Although not territorial, they tend to remain in one area.
REACTION TO DIVERS: Where spearfishing occurs, Kelp Bass are extremely wary and rapidly retreat when approached. Inside protective reserves they are often unafraid and curious and may allow a close view if approached with slow nonthreatening movements.
NOTE: Also commonly known as "Calico Bass."

DISTINCTIVE FEATURES: 1. Many dark spots on pale undercolor. 2. First two spines of foredorsal fin short; third spine distinctly longer.
DESCRIPTION: Often have darkish bars on sides and midbody stripe; belly pale. Tail square-cut.
ABUNDANCE & DISTRIBUTION: Occasional southern California; rare central California. Also south to Gulf of California.
HABITAT & BEHAVIOR: Inhabit sandy areas near reefs, rocky outcroppings and eelgrass beds. Occasionally rest on bottom propped up on fins.
REACTION TO DIVERS: Appear unafraid; usually allow close view when approached with slow nonthreatening movements.

Heavy Body/Large Lips

GULF GROUPER
Mycteroperca jordani
FAMILY:
Sea Bass – Serranidae

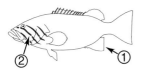

SIZE: 1½ - 3½ ft.,
max. 6½ ft.
DEPTH: 6 - 150 ft.

KELP BASS
Paralabrax clathratus
FAMILY:
Sea Bass – Serranidae

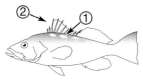

SIZE: 12-24 in.,
max. 28½ in.
DEPTH: 10 - 150 ft.

SPOTTED SAND BASS
Paralabrax maculatofasciatus
FAMILY:
Sea Bass – Serranidae

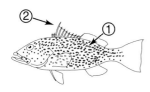

SIZE: 10 - 18 in.,
max. 22 in.
DEPTH: 5 - 200 ft.

Sea Bass – Rockfish

DISTINCTIVE FEATURES: 1. Dusky bars on side. (Distinguished from similar Spotted Sand Bass [previous] by lack of body spots.) **2. First two spines of foredorsal fin short; third spine distinctly longer.**
DESCRIPTION: Light gray, olive or brown upper body becomes lighter toward belly; usually small yellow-brown spots on head. Tail square-cut.
ABUNDANCE & DISTRIBUTION: Occasional southern California. Also south to Baja.
HABITAT & BEHAVIOR: Inhabit sandy areas near reefs, rocky outcroppings and kelp beds. Often rest on bottom propped up on pectoral fins. Occasionally in small groups; rarely deeper than safe diving limits.
REACTION TO DIVERS: Appear unafraid; usually allow close view when approached with slow nonthreatening movements.

DISTINCTIVE FEATURES: 1. Yellow speckles on head and body. 2. Yellow stripe from foredorsal fin curves to run length of lateral line to tail.
DESCRIPTION: Bluish black to black with yellow spots and blotches.
ABUNDANCE & DISTRIBUTION: Common southeast Alaska to California; occasional to rare northern and central California.
HABITAT & BEHAVIOR: Inhabit rocky inshore areas along exposed coastlines. Lurk in caves and crevices, resting on bottom propped up by their fins. Often remain in same "home site" for years. When away from hole, swim near bottom. Solitary.
REACTION TO DIVERS: Unafraid and curious; often allow close view when approached with slow nonthreatening movements.
NOTE: Also commonly known as "Yellowstripe Rockfish."

DISTINCTIVE FEATURES: 1. Bright yellow spots and blotches over black to dark olive brown undercolor. 2. Yellow membrane between third and fourth dorsal fin spines.
DESCRIPTION: Two dark diagonal bands extend from lower eye.
ABUNDANCE & DISTRIBUTION: Common central and southern California; occasional to uncommon northern California. Also south to central Baja.
HABITAT & BEHAVIOR: Inhabit rocky areas, lurking in caves and crevices. Often rest on bottom propped up by their fins. Territorial. Uncommon below 70 feet.
REACTION TO DIVERS: Unafraid and curious; often allow close view when approached with slow nonthreatening movements.

Heavy Body/Large Lips

BARRED SAND BASS
Paralabrax nebulifer
FAMILY:
Sea Bass – Serranidae

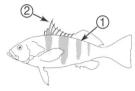

SIZE: 10-18 in.,
max. 26 in.
DEPTH: 5-600 ft.

CHINA ROCKFISH
Sebastes nebulosus
FAMILY:
Scorpionfish –
Scorpaenidae

SIZE: 8-14 in.,
max. 17 in.
DEPTH: 12-400 ft.

BLACK-AND-YELLOW ROCKFISH
Sebastes chrysomelas
FAMILY:
Scorpionfish –
Scorpaenidae

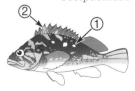

SIZE: 6-12 in.,
max. 15 1/4 in.
DEPTH: 0-150 ft.

Rockfish

DISTINCTIVE FEATURES: 1. Pale blotch extends onto back from between third and fourth dorsal spines, another at mid-foredorsal fin and a third where spinous and soft dorsal fins meet. 2. Several small white to pale yellow or reddish tan blotches on back.
DESCRIPTION: Reddish brown to greenish brown undercolor. Two diagonal bands extend from lower eye.
ABUNDANCE & DISTRIBUTION: Common central California; occasional to uncommon southern and northern California. Also south to central Baja.
HABITAT & BEHAVIOR: Inhabit rocky areas, lurking in caves and crevices. Often resting on bottom propped up by their fins. Territorial. Most common between 30-120 feet.
REACTION TO DIVERS: Unafraid and curious; often allow close view when approached with slow nonthreatening movements.

DISTINCTIVE FEATURES: 1. High spinous dorsal fin with deep notches between spines. 2. Large areas or blotches of white to yellow or orange on and below spinous dorsal fin.
DESCRIPTION: Orangish brown to dark brown or blackish; often pale area between eye and pectoral fin.
ABUNDANCE & DISTRIBUTION: Common southeastern Alaska to northern California; uncommon north to Gulf of Alaska and south to central California.
HABITAT & BEHAVIOR: Inhabit rocky offshore reefs and areas of large boulders; often in or near areas of kelp. Usually deeper than 70 feet in California, but as shallow as 30 feet in northern part of range. Gather in small schools near rocky bottoms, especially over boulder piles with numerous recesses. Remain in same area for years; seldom stray far except to attack prey such as schools of herring. Occasionally lurk in shaded protected recesses, resting on bottom propped up by their fins. Juveniles frequently seek shelter inside the hollow branches of large siliceous sponges. Most common rockfish in mainland inlets of British Columbia, especially over boulder-covered bottoms.
REACTION TO DIVERS: Wary, but often can be closely viewed when approached with slow nonthreatening movements.

DISTINCTIVE FEATURES: 1. Several irregular, rear slanting diagonal brown bands. (Similar Copper Rockfish [next] lack these markings.) 2. Whitish stripe runs along rear half of lateral line.
DESCRIPTION: Reddish white to pale tan, yellow or yellowish green undercolor with numerous small blotches and spots. Two dusky bands angle from eye toward pectoral fin.
ABUNDANCE & DISTRIBUTION: Common southern California; uncommon to rare central California. Also south to northern Baja.
HABITAT & BEHAVIOR: Inhabit flat soft bottoms, occasionally around rocky reefs and outcroppings.
REACTION TO DIVERS: Unafraid; usually allow close view when approached with slow nonthreatening movements.

Heavy Body/Large Lips

GOPHER ROCKFISH
Sebastes carnatus
FAMILY: Scorpionfish – Scorpaenidae

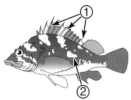

SIZE: 6-12 in., max. 15½ in.
DEPTH: 0-180 ft.

QUILLBACK ROCKFISH
Sebastes maliger
FAMILY: Scorpionfish – Scorpaenidae

SIZE: 6-18 in., max. 2 ft.
DEPTH: 5-900 ft.

CALICO ROCKFISH
Sebastes dalli
FAMILY: Scorpionfish – Scorpaenidae

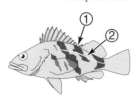

SIZE: 4-7 in., max. 10 in.
DEPTH: 60-900 ft.

Rockfish

DISTINCTIVE FEATURES: 1. Wide, whitish to pink stripe runs along rear half of lateral line. 2. Prominent dark band, often bordered with yellow, slopes downward from eye toward pectoral fin, another band below and yellow bar behind eye. (Similar Calico Rockfish [previous] has no yellow below or behind eye.)

DESCRIPTION: Patches of copper to yellow, orange, pink, brown or greenish brown over pale undercolor (tends to be copper to orange in northern range and yellow-brown to olive south of Monterey); belly white; fins pale, may be coppery or dusky. Frequently yellow bar behind eyes.

ABUNDANCE & DISTRIBUTION: Common southern British Columbia to southern California; uncommon north to Gulf of Alaska. Also south to Baja. Most common shallow-water rockfish in BC.

HABITAT & BEHAVIOR: Inhabit rocky areas from offshore reefs to shallow protected bays and inlets and areas of kelp; occasionally under docks and around jetties. Lurk in shadows and protected areas, often rest on bottom. Although not territorial, tend to remain in one area. Typically solitary, occasionally gather in small groups, hovering in kelp beds or above rocky bottoms.

REACTION TO DIVERS: Wary; often move away when approached.

NOTE: Also commonly known as "Whitebelly Rockfish" in southern range where it was formerly classified as a separate species *S. vexillaris*.

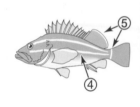

Yelloweye Rockfish Juvenile

DISTINCTIVE FEATURES: Red to orange-yellow. 1. Bright yellow iris. 2. Prominent spines between eyes and rough parallel ridges on nape. (Similar Redstripe Rockfish [next] distinguished by lack of spines and ridges.) 3. Prominent white stripe down lateral line of young that fades with age, becoming indistinct in full-sized adults near three feet. JUVENILE (to about one foot): 4. White stripe below lateral line. 5. Dorsal fin is edged in white, and a white bar at base of tail.

DESCRIPTION: Young are red, gradually changing to red-orange and to orange-yellow in full-sized adults.

ABUNDANCE & DISTRIBUTION: Occasional central and southern California; uncommon north to Gulf of Alaska. Also south to northern Baja.

HABITAT & BEHAVIOR: Inhabit rocky offshore reefs and cliff faces below 50 feet; large adults are usually deeper, often well below safe diving limits. Solitary, hide in deep crevices, caves and other protective recesses. In northern extent of range, juveniles often seek shelter inside the hollow branches of siliceous sponges.

REACTION TO DIVERS: Shy; move deep into recess when approached. After quiet, patient wait, often reappear to peer out of recess.

NOTE: Also commonly known as "Turkey-red Rockfish", "Rasphead" and incorrectly "Red

Heavy Body/Large Lips

COPPER ROCKFISH
Sebastes caurinus
FAMILY:
Scorpionfish –
Scorpaenidae

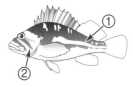

SIZE: 10 -16 in.,
max. 22½ in.
DEPTH: 0 - 600 ft.

Copper Rockfish
Yellowish brown variation, more common in southern extent of range.

YELLOWEYE ROCKFISH
Sebastes ruberrimus
FAMILY:
Scorpionfish –
Scorpaenidae

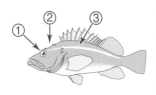

SIZE: 8 - 20 in.,
max. 3 ft.
DEPTH: 50 -1,200 ft.

Rockfish

DISTINCTIVE FEATURES: 1. Pale gray to pink stripe, bordered in darker red, runs length of lateral line from gill cover to tail. 2. Knob on tip of lower lip. (Similar Yelloweye Rockfish [previous] distinguished by lack of knob on lower lip and presence of prominent spines and rough ridges on snout and forehead.)
DESCRIPTION: Red to pinkish to orange back, becoming yellowish on sides; white belly.
ABUNDANCE & DISTRIBUTION: Rare (within safe diving limits) Bering Sea to Monterey Bay, California.
HABITAT & BEHAVIOR: Inhabit sand, mud and silt bottoms near rocky reefs, the base of cliffs and steep drop-offs. Rarely shallower than 90 feet. Little is known about the behavior of this deep-dwelling species.
REACTION TO DIVERS: Wary; move away when approached. Stalking with slow nonthreatening movements may be rewarded with close view.

DISTINCTIVE FEATURES: Shades of red to orange. **1. Soft dorsal, pectoral, ventral, anal and tail fins usually dark-edged.** (Similar Canary Rockfish [next] lack these markings.) **2. Thin pale stripe runs along lateral line from midbody to tail. 3. Band slopes downward from eye toward pectoral fin** (more prominent in smaller specimens) **with narrow band below.**
DESCRIPTION: Tends to be more reddish brown to orange in shallower water, changing to brilliant reds with depth; grayish undercolor. Ventral and anal fins rounded. (Fins of similar Canary Rockfish [next] are pointed.)
ABUNDANCE & DISTRIBUTION: Occasional California; uncommon north to Queen Charlotte Islands. Also south to central Baja.
HABITAT & BEHAVIOR: Inhabit rocky reefs and areas of large boulders and stones. Adults usually deeper (below 70 ft.); young shallower. May gather in small schools swimming near bottom.
REACTION TO DIVERS: Unafraid; usually allow close view when approached with slow nonthreatening movements.

Vermilion Rockfish
Bright red variations tend to be in deep water.

Heavy Body/Large Lips

REDSTRIPE ROCKFISH
Sebastes proriger
FAMILY:
Scorpionfish –
Scorpaenidae

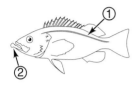

SIZE: 7-15 in.,
max. 20 in.
DEPTH: 60-1,200 ft.

VERMILION ROCKFISH
Sebastes miniatus
FAMILY:
Scorpionfish –
Scorpaenidae

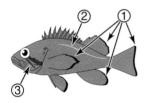

SIZE: 10-20 in.,
max. 30 in.
DEPTH: 40-900 ft.

Vermilion Rockfish
Reddish brown variations tend to be in shallow water.

Rockfish

DISTINCTIVE FEATURES: Bright orange to yellowish orange. **1. White stripe along lateral line from gill cover to tail. 2. Prominent band slopes downward from eye toward pectoral fin with narrow band below.**
DESCRIPTION: Grayish undercolor. Young (to 14 in.) have prominent dark spot at rear of spinous dorsal fin. Ventral and anal fins pointed. (Fins of similar Vermilion Rockfish [previous] rounded.)
ABUNDANCE & DISTRIBUTION: Common southeast Alaska to central California; occasional to uncommon north to western Gulf of Alaska and south to southern California. Also south to northern Baja.
HABITAT & BEHAVIOR: Inhabit rocky reefs and areas with large boulders and stones. Adults usually deeper than safe diving limits except in northern part of range where they often gather in small schools and swim near bottom along steep drop-offs. Young shallower and often hover in loose aggregations above bottom.
REACTION TO DIVERS: Wary; generally move away when approached. Stalking with slow nonthreatening movements may allow close view. Occasionally curious; schools may approach diver, but remain at a safe distance.

DISTINCTIVE FEATURES: Brownish orange to reddish orange. **1. Often dusky stripe runs from above pectoral fin down side toward base of tail. 2. Thin, pale lateral line.**
DESCRIPTION: Pinkish to white underside. Center of tail dusky.
ABUNDANCE & DISTRIBUTION: Common southern half of British Columbia and Puget Sound; occasional to rare north to Gulf of Alaska and south to northern California.
HABITAT & BEHAVIOR: Inhabit rough rocky areas with numerous caves, crevices and other protective recesses. Hover just above protective recesses, often in loose aggregations.
REACTION TO DIVERS: Wary; dart into protective recesses when approached. Slow nonthreatening movements may allow a close view.

DISTINCTIVE FEATURES: 1. A few, often vague, dark, squarish or rectangular blotches on back and across lateral line. 2. Pale bar, bordered by dark brown, angles from snout, under eye, and toward pectoral fin.
DESCRIPTION: Shades of light brown; whitish belly. Second spine of anal fin long.
ABUNDANCE & DISTRIBUTION: Abundant to common southern California; occasional to rare central California. Also south to central Baja.
HABITAT & BEHAVIOR: School in large numbers over high relief rocky reefs and outcroppings. Generally below 100 feet.
REACTION TO DIVERS: Appear unconcerned; generally allow a close view when approached with slow nonthreatening movements.

Heavy Body/Large Lips

CANARY ROCKFISH
Sebastes pinniger
FAMILY:
Scorpionfish –
Scorpaenidae

SIZE: 10 - 24 in.,
max. 30 in.
DEPTH: 30 - 900 ft.

PUGET SOUND ROCKFISH
Sebastes emphaeus
FAMILY:
Scorpionfish –
Scorpaenidae

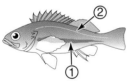

SIZE: 3 - 6 in.,
max. 7 in.
DEPTH: 30 - 1,200 ft.

SQUARESPOT ROCKFISH
Sebastes hopkinsi
FAMILY:
Scorpionfish –
Scorpaenidae

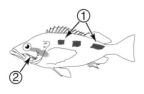

SIZE: 4 - 9 in.,
max. 11 1/2 in.
DEPTH: 50 - 600 ft.

Rockfish

DISTINCTIVE FEATURES: 1. Vague dusky to dark blotch on rear gill cover.
DESCRIPTION: Mottled and blotched shades of tan to coral to brown; pectoral and foredorsal fins usually pale coral to tan. (Similar Grass Rockfish [next] are distinguished by mottled and blotched shades of dark green to greenish gray or to nearly black; all fins dark.)
ABUNDANCE & DISTRIBUTION: Common to occasional Puget Sound to southern California; occasional to uncommon north to southeast Alaska. Also south to central Baja.
HABITAT & BEHAVIOR: Inhabit hard bottoms and sandy areas near rocks, dock pilings, debris and low profile reefs. Often lie quietly on bottom. More common below 20 feet.
REACTION TO DIVERS: Appear unconcerned; generally allow a close view when approached with slow nonthreatening movements.
NOTE: Also commonly known as "Bolinas."

DISTINCTIVE FEATURES: 1. Head and back mottled with dark blotches in shades of dark green to greenish gray or to nearly black on head and back. 2. Pectoral, dorsal and tail fins dark. (Similar Brown Rockfish [previous] distinguished by shades of brown; pectoral and foredorsal fins usually pale coral to tan.)
DESCRIPTION: Stout chunky body. Short, heavy foredorsal fin spines.
ABUNDANCE & DISTRIBUTION: Common California; uncommon north to central Oregon. Also south to central Baja.
HABITAT & BEHAVIOR: Wide range of shallow habitats from tide pools to eelgrass beds and rocky reefs and outcroppings with abundant plant life, including kelp; also around docks, piers and breakwaters. Most common between 2-30 feet.
REACTION TO DIVERS: Appear unconcerned; apparently relying on camouflage, move away only when closely approached.

DISTINCTIVE FEATURES: No distinctive markings.
DESCRIPTION: Mottled shades of tan to brown to greenish brown, occasionally with pinkish tints; can change color and markings to blend with background.
ABUNDANCE & DISTRIBUTION: Common southern and central California; occasional to uncommon northern California. Also south to central Baja.
HABITAT & BEHAVIOR: Inhabit kelp beds and areas of other algae. Drift in shaded areas of kelp forest, occasionally resting on blades or bottom. Most common between 20-60 feet.
REACTION TO DIVERS: Appear unconcerned; allow close approach before moving away.
NOTE: Also commonly known as "Dumb Bass."

Heavy Body/Large Lips

BROWN ROCKFISH
Sebastes auriculatus
FAMILY:
Scorpionfish –
Scorpaenidae

SIZE: 8 -18 in.,
max. 21½ in.
DEPTH: 0 - 400 ft.

GRASS ROCKFISH
Sebastes rastrelliger
FAMILY:
Scorpionfish –
Scorpaenidae

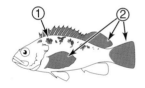

SIZE: 8 -16 in.,
max. 22 in.
DEPTH: 0 -150 ft.

KELP ROCKFISH
Sebastes atrovirens
FAMILY:
Scorpionfish –
Scorpaenidae

SIZE: 6 -14 in.,
max. 16¾ in.
DEPTH: 10 -150 ft.

Rockfish

DISTINCTIVE FEATURES: 1. One or two areas or streaks of yellow behind eye on gill cover. (Very similar Olive Rockfish [next] lack these markings; however, the first important clue to identification is location; the Olive is rare north of Point Conception while the Yellowtail is rare south.) **2. Several pale spots below dorsal fin.**
DESCRIPTION: Dark brown to green-brown or gray back, often pale below lateral line; reddish brown to orange-brown speckles on body scales (similar Olive Rockfish lack these speckles). Light green to yellow-green, yellow or dusky yellow fins. Rear edge of anal fin straight and vertical.
ABUNDANCE & DISTRIBUTION: Common to occasional Alaska to central California; rare southern California.
HABITAT & BEHAVIOR: Inhabit open water, especially over banks, reefs and along rapidly descending coastlines. Congregate in compact schools, often with other rockfishes, orienting into the prevailing current. Occasionally individuals may lurk near recesses in the bottom. At night commonly rest on bottom.
REACTION TO DIVERS: Wary; retreat when approached. Following with slow nonthreatening movements may allow close view.

DISTINCTIVE FEATURES: 1. Several pale spots below dorsal fin. (Very similar Yellowtail [previous] has one or two streaks of yellow behind eye; however, the first important clue to identification is location; the Olive is rare north of Point Conception while the Yellowtail is rare south.)
DESCRIPTION: Mottled shades greenish brown, lighter below lateral line; fins greenish brown to dusky yellow.
ABUNDANCE & DISTRIBUTION: Common southern California; uncommon to rare central California; rare northern California. Also south to central Baja.
HABITAT & BEHAVIOR: Inhabit open water, especially over banks, reefs and along rapidly descending coastlines. Most common between 20-120 feet. Congregate in schools, may mix with Blue Rockfish. Individuals occasionally rest on bottom.
REACTION TO DIVERS: Wary; retreat when approached. Often can be closely observed by following with slow nonthreatening movements.

DISTINCTIVE FEATURES: 1. Dark membrane between pale rays of pectoral, ventral and anal fins. 2. Often display large, pale blotch on side.
DESCRIPTION: Shades of brown, often with yellow tints; belly white, occasionally with reddish wash. Forehead concave between eyes. Rear edge of anal fin relatively straight and slanting.
ABUNDANCE & DISTRIBUTION: Common to occasional Alaska to southern California. Also to southern Baja.
HABITAT & BEHAVIOR: Inhabit offshore reefs and banks. Generally larger individuals remain below 80 feet; smaller specimens in shallower water where young tend to school. Actively swim in open water above bottom; occasionally gather into huge dense schools.
REACTION TO DIVERS: Wary; move away when approached. Stalking with slow nonthreatening movements may allow a close view.

Heavy Body/Large Lips

YELLOWTAIL ROCKFISH
Sebastes flavidus
FAMILY:
Scorpionfish –
Scorpaenidae

SIZE: 12-22 in.,
max. 26 in.
DEPTH: 0 - 900 ft.

OLIVE ROCKFISH
Sebastes serranoides
FAMILY:
Scorpionfish –
Scorpaenidae

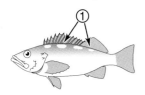

SIZE: 10 - 20 in.,
max. 2 ft.
DEPTH: 10 - 500 ft.

WIDOW ROCKFISH
Sebastes entomelas
FAMILY:
Scorpionfish –
Scorpaenidae

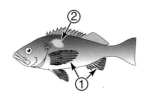

SIZE: 8 -17 in.,
max. 21 in.
DEPTH: 0 -1,200 ft.

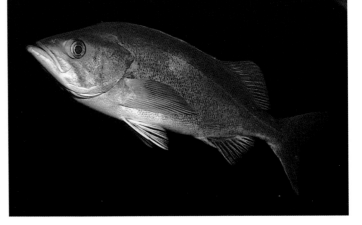

Rockfish

DISTINCTIVE FEATURES: 1. Large mouth with projecting lower jaw and prominent knob at tip. 2. Upper jaw extends to rear of eye. (Similar Widow Rockfish [previous] distinguished by small mouth and only slightly extended lower jaw. Very similar Bocaccio [next] has slightly larger lower lip, more upturned mouth and upper jaw extends past eye.)

DESCRIPTION: Gray or brown back changing to silver-gray or tan to reddish tan sides, often with greenish tints, and white belly. (Similar Bocaccio [next] often has dark specks on back and body and silver belly.) Forehead profile concave. Rear edge of anal fin somewhat rounded.

ABUNDANCE & DISTRIBUTION: Uncommon Bering Sea to Oregon; rare south to southern California.

HABITAT & BEHAVIOR: Inhabit wide range of habitats. Solitary; slowly cruise just above the bottom or lurk in caves, crevices and other recesses. Generally, only smaller individuals venture into shallow waters; large specimens usually remain below safe diving limits.

REACTION TO DIVERS: Wary; move away or dart into protective recesses when approached. A slow nonthreatening approach may allow a closer view.

DISTINCTIVE FEATURES: 1. Large, upturned mouth with projecting lower jaw and prominent knob at tip. 2. Upper jaw extends past eye. (Similar Silvergray Rockfish [previous] has less upturned mouth and upper jaw extends to only rear of eye.)

DESCRIPTION: Dark gray to silvery gray, reddish gray, or olive-brown to olive gradating to silvery belly; often with scattering of dark spots, especially on young. (Silvergray Rockfish lack these spots.) Head profile concave.

ABUNDANCE & DISTRIBUTION: Common California; occasional Oregon and Washington; uncommon to rare north to Alaska. Also south to central Baja.

HABITAT & BEHAVIOR: Inhabit wide range of habitats. Slowly cruise just above the bottom. Generally, only smaller individuals inhabit shallow depths; large specimens usually remain below safe diving limits. Occasionally large individuals venture into shallower depths, especially in areas with cold upwelling currents.

REACTION TO DIVERS: Wary; move away or dart into protective recesses when approached. A slow nonthreatening approach may allow a closer view.

Bocaccio

Reddish brown color variation with scattering of dark spots that is especially common on smaller individuals.

Heavy Body/Large Lips

SILVERGRAY ROCKFISH
Sebastes brevispinis

FAMILY:
Scorpionfish –
Scorpaenidae

SIZE: 12-22 in.,
max. 28 in.
DEPTH: 0 -1,200 ft.

BOCACCIO
Sebastes paucispinis

FAMILY:
Scorpionfish –
Scorpaenidae

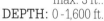

SIZE: 8 -28 in.,
max. 3 ft..
DEPTH: 0 -1,600 ft.

Bocaccio
Dark gray color variation typical of large individuals; note large knob at tip of lower lip.

41

Rockfish

DISTINCTIVE FEATURES: Lightly mottled gray to black. **1. Moderate knob at tip of slightly projecting lower jaw. 2. Forehead profile slightly convex.** (Similar Silvergray and Widow Rockfish and Bocaccio [previous] distinguished by concave head profile.) **3. Rear edge of anal fin straight and relatively vertical.** (Similar Black Rockfish [next] distinguished by rounded anal fin; Blue Rockfish [bottom of page] distinguished by straight, but slanting, rear edge of anal fin.)
DESCRIPTION: Occasionally tinted with brown, blue or green.
ABUNDANCE & DISTRIBUTION: Uncommon Bering Sea to central British Columbia and northern Vancouver Island.
HABITAT & BEHAVIOR: Inhabit offshore reefs and rocky shorelines. Often rest alone on bottom propped up on fins or hover just above bottom in loose aggregations.
REACTION TO DIVERS: Wary; move toward shelter of crevices or other recesses when approached. Occasionally, when resting on bottom, may be approached with slow nonthreatening movements.

DISTINCTIVE FEATURES: Mottled and blotched with black to blue-black and gray. **1. Small knob at tip of slightly projecting lower jaw that extends to or past eye.** (Similar Blue Rockfish [next] jaw extends only to mideye.) **2. Rear edge of anal fin rounded.** (Similar Dusky Rockfish [previous] and Blue Rockfish [next] distinguished by straight-edged rear anal fin.)
DESCRIPTION: Back generally dark, often lighter shades below lateral line. Forehead profile convex.
ABUNDANCE & DISTRIBUTION: Abundant to occasional Aleutian chain of Alaska to southern California.
HABITAT & BEHAVIOR: Inhabit wide range of habitats from kelp forests to rocky, boulder-strewn bottoms and in open water over deep banks. Often gather in huge schools, frequently with other rockfishes, especially Blue Rockfish. Aggressive feeders, often churning surface waters when hunting as a school. Commonly rest on bottom at night.
REACTION TO DIVERS: Wary; move away when approached. When resting on bottom, may be closely viewed when approached with slow nonthreatening movements.

DISTINCTIVE FEATURES: Mottled blue-black to bright blue. **1. Two to four dark bands curve around front of head. 2. Sloping band from the eye toward the pectoral fin with a smaller band below. 3. Rear edge of anal fin straight and slanted.** (Similar Dusky Rockfish [top of page] distinguished by relatively vertical straightedge, and Black Rockfish [previous] by rounded anal fin.)
DESCRIPTION: Back dark, generally paler below; fins blackish. Slightly projecting jaw extends only to midpoint of eye. (Similar Black Rockfish [previous] jaw extends to or past rear of eye.) Forehead profile slightly convex.
ABUNDANCE & DISTRIBUTION: Common central California north to central British Columbia and Vancouver Island; uncommon southern California; rare north to Bering Sea. Also south to northern Baja.
HABITAT & BEHAVIOR: Inhabit kelp forests, shallow reefs and open water over deep reefs, rarely in sheltered waters. Form large schools with other rockfishes, most commonly mixed in with Black Rockfish. Aggressive feeders, often churning surface waters when hunting as a school. Commonly rest on bottom at night.
REACTION TO DIVERS: Wary, but curious; generally move away when approached. After a short time fish may approach a diver who remains motionless.

Heavy Body/Large Lips

DUSKY ROCKFISH
Sebastes ciliatus
FAMILY:
Scorpionfish –
Scorpaenidae

SIZE: 6-12 in.,
max. 16 in.
DEPTH: 0-900 ft.

BLACK ROCKFISH
Sebastes melanops
FAMILY:
Scorpionfish –
Scorpaenidae

SIZE: 8-18 in.,
max. 2 ft.
DEPTH: 0-1,200 ft.

BLUE ROCKFISH
Sebastes mystinus
FAMILY:
Scorpionfish –
Scorpaenidae

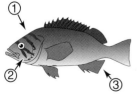

SIZE: 8-18 in.,
max. 21 in.
DEPTH: 0-300 ft.

Rockfish

DISTINCTIVE FEATURES: 1. Several large white spots on back. 2. Outlined scales below lateral line form a honeycomb pattern. (Similar Rosy Rockfish [next] does not have honeycomb pattern; Starry Rockfish [bottom of page] distinguished by numerous, small white spots.)

DESCRIPTION: Pinkish tan to orange or yellow; back darker.

ABUNDANCE & DISTRIBUTION: Common southern California; rare central California. Also south to central Baja.

HABITAT & BEHAVIOR: Inhabit deep, offshore rocky reefs. Usually drift just above bottom, occasionally rest on substrate.

REACTION TO DIVERS: Unafraid; can usually be closely approached with slow nonthreatening movements.

DISTINCTIVE FEATURES: 1. Several large white spots bordered in rose to dark red or purple on back. (Similar Honeycomb Rockfish [previous] distinguished by honeycomb markings on sides; Starry Rockfish [next] distinguished by numerous, small white spots.) **2. Rose to red or purplish area between eyes.**

DESCRIPTION: Rose to red or purplish blotches and/or mottling over red to orange undercolor; fins purplish red to orange.

ABUNDANCE & DISTRIBUTION: Occasional California. Also south to central Baja.

HABITAT & BEHAVIOR: Inhabit deep, offshore rocky, high relief reefs, often in caves, crevices or under overhangs. Usually drift above bottom. Young rarely above 90 feet and adults are usually below safe diving limits.

REACTION TO DIVERS: Unafraid; can usually be closely approached with slow nonthreatening movements.

DISTINCTIVE FEATURES: 1. Numerous small, white spots cover body. (Similar Honeycomb Rockfish [top of page] distinguished by honeycomb pattern on sides. Similar Rosy Rockfish [previous] distinguished by lack of small white spots.) **2. Several large white spots on back.**

DESCRIPTION: Red to orange; back darker.

ABUNDANCE & DISTRIBUTION: Common southern California; occasional to rare north to San Francisco. Also south to southern Baja.

HABITAT & BEHAVIOR: Inhabit deep, offshore rocky reefs. Drift, or rest on substrate; frequently in caves, crevices and recesses, occasionally in open just above bottom.

REACTION TO DIVERS: Unafraid; can usually be closely approached with slow nonthreatening movements.

Heavy Body/Large Lips

HONEYCOMB ROCKFISH
Sebastes umbrosus
FAMILY:
Scorpionfish –
Scorpaenidae

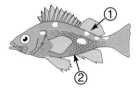

SIZE: 4 - 8 in.,
max. 10 1/2 in.
DEPTH: 90 - 400 ft.

ROSY ROCKFISH
Sebastes rosaceus
FAMILY:
Scorpionfish –
Scorpaenidae

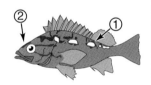

SIZE: 5 - 10 in.,
max. 14 in.
DEPTH: 50 - 450 ft.

STARRY ROCKFISH
Sebastes constellatus
FAMILY:
Scorpionfish –
Scorpaenidae

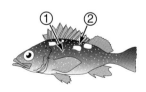

SIZE: 6 - 15 in.,
max. 1 1/2 ft.
DEPTH: 80 - 900 ft.

Rockfish

DISTINCTIVE FEATURES: 1. Five dark red to black body bars, no bars on base of tail. (Similar Treefish [next] distinguished by bar on base of tail; Flag Rockfish [next page] distinguished by bar at base of tail and another on tail; Calico Rockfish [pg. 29] has diagonal bands rather than bars.)
DESCRIPTION: Pink to coral or red. Two dark bands angle from eye toward pectoral fin and often a short band extends from eye toward back.
ABUNDANCE & DISTRIBUTION: Occasional British Columbia and Vancouver Island; uncommon to rare north to Alaska and south to central California.
HABITAT & BEHAVIOR: Inhabit areas with numerous caves, crevices and other protective recesses where they lurk, often hidden from view. Solitary and territorial.
REACTION TO DIVERS: Usually shy; retreat into recess when approached. Occasionally aggressively territorial, erecting their large, formidable spinous dorsal fins to challenge intruders.

DISTINCTIVE FEATURES: Bright to dusky yellow **1. Five to six wide (occasionally double) black to dark olive body bars across back and base of tail.** (Yellow undercolor easily distinguishes this species from other barred rockfishes.)
DESCRIPTION: Two dark bands angle from eye toward pectoral fin. Pink lips.
ABUNDANCE & DISTRIBUTION: Common to occasional southern California; uncommon to rare north to San Francisco. Also south to central Baja.
HABITAT & BEHAVIOR: Inhabit areas with numerous caves, crevices and other protective recesses where they lurk, often hidden from view. Solitary and territorial. Most common between 20-100 feet.
REACTION TO DIVERS: Usually shy; retreat into recess when approached. Occasionally aggressively territorial, erecting their large, formidable spinous dorsal fins to challenge intruders.

DESCRIPTION: Color and markings are more intense. **2. Fins edged with white to iridescent blue.**

Heavy Body/Large Lips

TIGER ROCKFISH
Sebastes nigrocinctus
FAMILY:
Scorpionfish –
Scorpaenidae

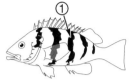

SIZE: 8-20 in.,
max. 2 ft.
DEPTH: 30-900 ft.

TREEFISH
Sebastes serriceps
FAMILY:
Scorpionfish –
Scorpaenidae

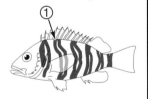

SIZE: 6-14 in.,
max. 16 in.
DEPTH: 10-150 ft.

Treefish Young

Rockfish

DISTINCTIVE FEATURES: 1. Four red to red-orange to reddish brown body bars across back and base of tail. 2. Submarginal bar on tail fin.
DESCRIPTION: White to pinkish white or ivory. Three prominent bands radiate from eye. First and second body bars are curved toward rear. Bars darker in young.
ABUNDANCE & DISTRIBUTION: Common to occasional southern and central California; uncommon to rare to San Francisco. Also south to northern Baja.
HABITAT & BEHAVIOR: Inhabit wide range of hard-bottom habitats. Often shelter in and around rocks, large anemones, ledge overhangs, entrance to protective recesses and kelp. Generally solitary. Young shallower; large adults usually below safe diving limits.
REACTION TO DIVERS: Usually shy and retreat into recess when approached. Occasionally can be approached with slow nonthreatening movements.

DISTINCTIVE FEATURES: 1. Pinkish stripe, bordered with two, often irregular, green to greenish brown stripes, runs down lateral line.
DESCRIPTION: Green to greenish brown or brown spots and blotches often form irregular stripes over pinkish undercolor.
ABUNDANCE & DISTRIBUTION: Rare (common below safe diving limits) Alaska to southern California. Also south to central Baja.
HABITAT & BEHAVIOR: Inhabit wide range of habitats from sandy or other soft bottoms to areas of mixed boulders and rocks. May form small aggregations near crevices.
REACTION TO DIVERS: Wary; when approached, move away, keeping a safe distance. Occasionally allow close view with slow nonthreatening movements.
NOTE: Also commonly known as "Strawberry Rockfish" and "Poinsettia Rockfish."

Heavy Body/Large Lips

FLAG ROCKFISH
Sebastes rubrivinctus

FAMILY:
Scorpionfish –
Scorpaenidae

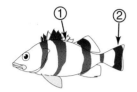

SIZE: 4 - 16 in.,
max. 20 in.
DEPTH: 100 - 600 ft.

GREENSTRIPED ROCKFISH
Sebastes elongatus

FAMILY:
Scorpionfish –
Scorpaenidae

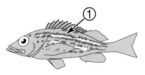

SIZE: 5 - 10 in.,
max. 15 in.
DEPTH: 100 - 1,400 ft.

IDENTIFICATION GROUP 2
Bulbous, Spiny-Headed Bottom-Dwellers
Scorpionfish – Sculpin

This ID Group consists of fish with large, bulbous heads, usually with numerous spines and/or skin flaps, and large, somewhat protruding eyes. They normally rest on the bottom blending with the background.

FAMILY: Scorpionfish — Scorpaenidae
4 Species Included

Scorpionfish
(typical shape)

Worldwide, members of the scorpionfish family vary from the beautifully gaudy lionfish and leaffish of the tropical Indo-Pacific to the rather drab California Scorpionfish. In addition to the rockfishes, discussed in the previous chapter, there are four additional scorpionfish that inhabit the waters of our region. Family members have venomous spines in their foredorsal, ventral and anal fins. Puncture wounds inflicted by local species are painful, but seldom, if ever, life threatening. Wounds should be: (1) immersed in nonscalding hot water (about 110 degrees) for 30-90 minutes, (2) then scrubbed with soap and water, (3) and irrigated with fresh water; (4) if the wound shows signs of infection, contact a physician.

Scoripionfish are masters of camouflage that can change their color and the shape of their skin flaps to blend with the substrate. Concealed, they rest motionless on the bottom waiting for the unsuspecting prey to pass near their large mouths.

Although each of the four species is similar in physical appearance, distinctive markings make visual identification possible.

FAMILY: Sculpin — Cottidae
27 Species Included

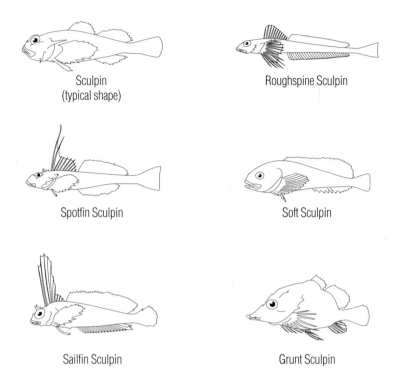

Sculpin (typical shape)

Roughspine Sculpin

Spotfin Sculpin

Soft Sculpin

Sailfin Sculpin

Grunt Sculpin

Sculpins are a large family of cold-water, bottom-dwelling fish. The majority have large bulbous heads that often bear fleshy appendages and numerous spines, especially on the gill covers. Protruding eyes set high on the head provide a wide range of vision. Nearly all sculpins have large pectoral fins and noticeable scales or bony tubes embedded along the lateral line. Shades of brown, green and gray predominate in this family. While most are only a few inches long, a few, including the Cabezon, grow much larger.

Sculpin typically live in tide pools and other intertidal habitats, but several species inhabit deeper water. It takes a keen eye to spot motionless sculpins blending with the bottom. When they do move, they tend to hug the sea floor.

Identification is generally based on subtle, but noticeable, differences. Most are only partly scaled and the patterns formed by these scaled areas are often used to identify the genus. Their ability to change color and markings often hampers identification.

Scorpionfish

DISTINCTIVE FEATURES: 1. Numerous brown spots on head, body and fins. (Similar Stone Scorpionfish [next] lack these spots.) **2. Numerous spines and short barbels and skin flaps on head.**
DESCRIPTION: Wide range of mottled and blotched shades from red to reddish brown, brown, tan and occasionally lavender. Can pale, darken or change color to blend with background.
ABUNDANCE & DISTRIBUTION: Common southern California; occasional to uncommon north to central California. Also south to Baja and Gulf of California.
HABITAT & BEHAVIOR: Inhabit recesses on rocky reefs. Lie on bottom blending with background; often nestled in with debris. Most commonly from 20 feet to well below safe diving limits.
REACTION TO DIVERS: Remain still, apparently relying on camouflage. Move only when closely approached or molested. Venomous dorsal fin spines can cause a painful wound.
NOTE: Also commonly known as "Spotted Scorpionfish."

California Scorpionfish
Color variation.

DISTINCTIVE FEATURES: 1. Two dark bars on tail and a third on tail base. 2. Often numerous, conspicuous barbels under mouth. (Similar California Scorpionfish [previous] distinguished by numerous spots on body.)
DESCRIPTION: Broad head and body, often with numerous skin flaps. Mottled with blotches in earthtones, and occasionally include shades of red and lavender. Can pale, darken or change color to blend with background. Occasionally have numerous white spots.
ABUNDANCE & DISTRIBUTION: Rare southern California. Also south to Ecuador and Galapagos Islands.
HABITAT & BEHAVIOR: Inhabit rocky, boulder and gravel-strewn slopes and on ledges along walls.
REACTION TO DIVERS: Remain still, apparently relying on camouflage. Move only when closely approached or molested. Venomous dorsal fin spines can cause a painful wound.
NOTE: Also commonly known as "Spotted Scorpionfish." Previously classified as *S. plumieri mystes* — a form of the Atlantic population *S. plumieri*.

Bulbous, Spiny-Headed Bottom-Dwellers

CALIFORNIA SCORPIONFISH
Scorpaena guttata
FAMILY:
Scorpionfish –
Scorpaenidae

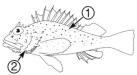

SIZE: 7-14 in.,
max. 17 in.
DEPTH: 0-600 ft.

California Scorpionfish
Color variation.

STONE SCORPIONFISH
Scorpaena mystes
FAMILY:
Scorpionfish –
Scorpaenidae

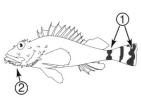

SIZE: 8-14 in.,
max. 1½ ft.
DEPTH: 1-100 ft.

Scorpionfish – Sculpin

DISTINCTIVE FEATURES: 1. Prominent dark spot on lower rear edge of gill plate.
DESCRIPTION: Mottled and blotched in shades of red to reddish brown and white. Can pale, darken or change color to blend with background. No barbels under mouth.
ABUNDANCE & DISTRIBUTION: Uncommon to rare southern California. Also south to Peru.
HABITAT & BEHAVIOR: Inhabit ledge overhangs and in other recesses on rocky reefs, steep slopes, and especially on walls. Small individuals often near protective spines of sea urchins. More common 20-60 feet.
REACTION TO DIVERS: Remain still, apparently relying on camouflage. Move only when closely approached or molested. Venomous dorsal fin spines can cause a painful wound.

DISTINCTIVE FEATURES: Red to pink. 1. Numerous spines on head.
DESCRIPTION: Often a scattering of dark blotches; pale underside. Pectoral fin bilobed; fourth or fifth spine of foredorsal fin longest.
ABUNDANCE & DISTRIBUTION: Rare (common below safe diving limits) Bering Sea to southern California. Also south to northern Baja.
HABITAT & BEHAVIOR: Inhabit soft sand and muddy bottoms. Lie on bottom, often nestled in with debris. Young more common in shallow depths.
REACTION TO DIVERS: Remain still, apparently relying on camouflage. Move only when closely approached or molested. Venomous dorsal fin spines can cause a painful wound.

DISTINCTIVE FEATURES: 1. Broad head with large mouth that extends beyond eye. (Similar Buffalo Sculpin [pg. 59] distinguished by short, steep snout and mouth that only extends to mid-eye.) 2. Long smooth spine extends from upper cheek.
DESCRIPTION: Variable patches, saddles and blotches of white to earthtones to nearly uniform brown to gray; occasionally display spots and small blotches of yellow (pictured); fins usually banded. Can change color and markings to blend with background. Bumpy, fleshy papillae on head contain embedded scales; body unscaled. Spinous and soft dorsal fins separated. (Similar Red and Brown Irish Lords [pg. 61] distinguished by notched, but continuous, dorsal fin.)
ABUNDANCE & DISTRIBUTION: Occasional to uncommon northern Washington and Puget Sound and north to Bering Sea and Aleutian Islands, Alaska. Also to northern Japan.
HABITAT & BEHAVIOR: Inhabit sandy, silty and rubble-strewn bottoms and occasionally rocky substrates. Often in the vicinity of piers, wharves and jetties. Rest on bottom, blending with surroundings.
REACTION TO DIVERS: Remain still, apparently relying on camouflage. Bolt only when closely approached or molested.

Bulbous, Spiny-Headed Bottom-Dwellers

RAINBOW SCORPIONFISH
Scorpaenodes xyris
FAMILY:
Scorpionfish –
Scorpaenidae

SIZE: 1-4 1/2 in.,
max. 6 in.
DEPTH: 0-100 ft.

SHORTSPINE THORNYHEAD
Sebastolobus alascanus
FAMILY:
Scorpionfish –
Scorpaenidae

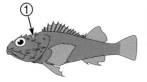

SIZE: 6-18 in.,
max. 29 1/2 in.
DEPTH: 55-5,000 ft.

GREAT SCULPIN
Myoxocephalus polyacanthocephalus
FAMILY:
Sculpin – Cottidae

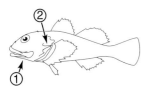

SIZE: 16-30 in.,
max. 30 in.
DEPTH: 0-130 ft.

Sculpin

DISTINCTIVE FEATURES: 1. Short cirrus centered near tip of snout. 2. Prominent cirrus above each eye. 3. Notch in spinous dorsal fin after third or fourth spine.

DESCRIPTION: Marbled earthtones; males generally display red shades, while females are usually greenish. Can change color, pale or darken to blend with background. Unscaled bulbous head and stout body. **JUVENILES:** Prominent cirrus centered on snout.

ABUNDANCE & DISTRIBUTION: Common southern California and north to southern British Columbia; occasional to uncommon to southeastern Alaska. Also to central Baja.

HABITAT & BEHAVIOR: Inhabit rocky bottoms, especially near kelp beds; often along exposed coasts and in tidal passages. Rest on bottom camouflaged with background. Mate in late winter, couples use same nesting site year after year. Females lay 50,000 to 100,000 purple to blue-green, pink or white eggs (note photograph). Males guard nest from predators, bolting only when closely approached, but return as soon as immediate danger passes. Consequently, they are unusually vulnerable to spearfishing which threatens future populations.

REACTION TO DIVERS: Remain still, apparently relying on camouflage. Bolt only when closely approached.

DISTINCTIVE FEATURES: 1. Long, branching spine extends from gill cover. 2. Dark blotch toward rear of first dorsal fin. 3. Dark to dusky bars on pectoral, dorsal and tail fins.

DESCRIPTION: Drab back spotted, blotched and mottled earthtone shades with greenish tints or gray; lower sides and belly cream to white; fins yellowish. Can change color and markings to blend with substrate. Spinous and soft dorsal fins separated. No scales.

ABUNDANCE & DISTRIBUTION: Common to occasional Bering Sea to southern California; however, seldom observed because of cryptic nature. Also south to central Baja.

HABITAT & BEHAVIOR: Inhabit inshore mud, silt, sand and shell rubble bottoms, especially in inlets, bays and estuaries. Rest on bottom, often bury with only upper head and eyes exposed. Most common in shallow waters; also inhabit depths below safe diving limits.

REACTION TO DIVERS: Remain still, apparently relying on camouflage. Bolt only when closely approached or molested; then generally move only a short distance and re-bury.

Bulbous, Spiny-Headed Bottom-Dwellers

CABEZON
Scorpaenichthys marmoratus
FAMILY:
Sculpin – Cottidae

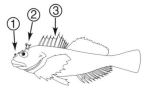

SIZE: 16-30 in., max. 39 in.
DEPTH: 0-250 ft.

Cabezon Juvenile
[right]
Note prominent single cirrus centered on snout.

Color variation
[far left]

Male
Guarding eggs.
[near left]

PACIFIC STAGHORN SCULPIN
Leptocottus armatus
FAMILY:
Sculpin – Cottidae

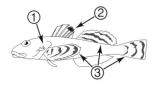

SIZE: 5-14 in., max. 18 in.
DEPTH: 0-300 ft.

Sculpin

DISTINCTIVE FEATURES: 1. Wide head with short, steep snout and mouth that extends to mid-eye. (Similar Great Sculpin [pg. 55] distinguished by long snout and mouth that extends beyond the eye.) **2. Long smooth spine extends from upper cheek with smaller spine below. 3. Lateral line high on back with large, raised scales (remainder of body unscaled).**

DESCRIPTION: Drab mottled and blotched shades of earthtones, often greenish gray to greenish or reddish brown; fins usually banded. Can change color and markings to blend with background. Spinous and soft dorsal fins separated. (Similar Red and Brown Irish Lords [pg. 61] distinguished by notched, but continuous, dorsal fin.)

ABUNDANCE & DISTRIBUTION: Occasional Alaska to central California. May be locally common or even abundant, but seldom observed because of cryptic nature.

HABITAT & BEHAVIOR: Inhabit shallow rocky areas mixed with algae and sandy areas often mixed with rubble. Rest on bottom, blending with surroundings. From February to March females lay small clusters of orange-brown eggs which the males guard and oxygenate by fanning with their pectoral fins.

REACTION TO DIVERS: Remain still, apparently relying on camouflage. Bolt only when closely approached or molested. Males usually remain when guarding eggs.

Buffalo Sculpin
Reddish brown color variation.

DISTINCTIVE FEATURES: 1. Erect spines on bulbous head. 2. Spinous and soft dorsal fins rounded and separated by deep notch.

DESCRIPTION: Head and back mottled and blotched in shades of gray to pinkish gray, often with brown spots; become brownish on sides to whitish underside. Often dark saddles on back, and fins with dark, irregular bands. Hair-like cirri on head and body. Short inconspicuous spine on cheek.

ABUNDANCE & DISTRIBUTION: Uncommon Aleutian Islands to Puget Sound and northern Washington; also to central Japan.

HABITAT & BEHAVIOR: Inhabit sand, silt and mud bottoms, especially where scattered with rocks, rubble and outcroppings. Rest on bottom, blending with surroundings.

REACTION TO DIVERS: Remain still, apparently relying on camouflage. Bolt only when closely approached or molested.

Bulbous, Spiny-Headed Bottom-Dwellers

BUFFALO SCULPIN
Enophrys bison
FAMILY:
Sculpin – Cottidae

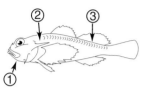

SIZE: 5-12 in., max. 14 1/2 in.
DEPTH: 0-65 ft.

Buffalo Sculpin
Male guarding cluster of orange eggs; red color variation.

SPINYHEAD SCULPIN
Dasycottus setiger
FAMILY:
Sculpin – Cottidae

SIZE: 4-7 in., max. 9 in.
DEPTH: 50-450 ft.

Sculpin

DISTINCTIVE FEATURES: 1. Continuous dorsal fin notched after third spine and again between spinous and soft rays. (Similar Great and Buffalo Sculpins [pgs. 55 & 59] distinguished by separate spinous and soft dorsal fins.) **2. Nostrils do not have fleshy flaps. 3. Stripe of scales four or five wide below dorsal fin, followed by an unscaled area to lateral line.**
DESCRIPTION: Mottled, spotted and blotched in shades of red to pink, red-brown, brown and white; usually four dark saddle-like bars on back. Can change color and markings to blend with background.
ABUNDANCE & DISTRIBUTION: Common Bering Sea, Alaska to Washington; rare to central California.
HABITAT & BEHAVIOR: Inhabit shallow rocky areas and reefs mixed with anemones, soft corals, pink coralline algae and other bottom growth. Rest on bottom, blending with surroundings. During winter until March females lay large masses of pink to purple to blue or yellow eggs which the males guard.
REACTION TO DIVERS: Remain still, apparently relying on camouflage. Bolt only when closely approached or molested. Males usually remain when guarding eggs.
SIMILAR SPECIES: Brown Irish Lord, *H. spinosus*, distinguished by nostrils that have fleshy flaps and a stripe of scales six to eight wide; brown coloration, never shades of red or pink. Uncommon to rare Southeast Alaska to southern California.

Red Irish Lord
Dark color variation blending with rocky substrate.

DISTINCTIVE FEATURES: 1. No cirri between eyes. 2. Almost completely scaled between dorsal fin and lateral line. (Similar Scalyhead Sculpin [next] has only a wide band of scales between dorsal fin and lateral line and has cirri between eyes.)
DESCRIPTION: Mottled, spotted and blotched earthtone shades, usually with several dark saddle markings on back extending to lateral line, with paler shades below; occasionally have ocellated markings on foredorsal fin (pictured). Can change color and markings to blend with background. Star-like scales on head; embedded scales along lateral line have thread-like extensions.
ABUNDANCE & DISTRIBUTION: Abundant to common Aleutian Islands, Alaska to Washington; occasional south to southern California.
HABITAT & BEHAVIOR: Inhabit wide range of shallow habitats from sandy rubble-strewn bottoms to rocky coastlines and areas of eelgrass; most common in less than 20 feet. Often among barnacles and mussels attached to wharf pilings and artificial reefs.
REACTION TO DIVERS: Remain still; bolt only when closely approached.
SIMILAR SPECIES: Smoothhead Sculpin, *A. lateralis*, distinguished by somewhat flattened snout, narrow band of scales below dorsal fin, and all fins (except ventral) barred.

Bulbous, Spiny-Headed Bottom-Dwellers

RED IRISH LORD
Hemilepidotus hemilepidotus
FAMILY:
Sculpin – Cottidae

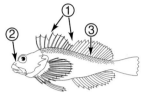

SIZE: 8-16 in.,
max. 20 in.
DEPTH: 0-160 ft.

Red Irish Lord
Brownish color variation is often mistaken for similar Brown Irish Lord.

PADDED SCULPIN
Artedius fenestralis
FAMILY:
Sculpin – Cottidae

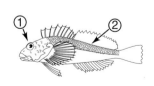

SIZE: 3-4 in.,
max. 5½ in.
DEPTH: 0-180 ft.

Sculpin

DISTINCTIVE FEATURES: 1. Red to reddish markings, often encircle a dark spot, on first two dorsal fin segments. 2. Broad band of scales between dorsal fin and lateral line. 3. Small cirri on snout and larger ones between eyes, especially large on males. (Similar Padded Sculpin [previous] is almost entirely scaled between dorsal fin and lateral line and has no cirri between eyes. Similar Coralline Sculpin [next] distinguished by less pronounced cirri and a longer and more pointed snout.)

DESCRIPTION: Mottled, spotted and blotched earthtone shades, often with red; usually several saddle markings on back to lateral line; pale to white spots below becoming entirely white on belly. Can change color and markings to blend with background. Flattened flesh-covered spine on cheek with two points. Have scales on head; embedded scales along lateral line have thread-like extensions.

ABUNDANCE & DISTRIBUTION: Abundant to common Aleutian Islands, Alaska, to Washington; occasional to rare south to southern California.

HABITAT & BEHAVIOR: Inhabit shallow reefs, rocky outcroppings and on dock pilings. Most common shallower than 20 feet. Unlike most sculpins, actively move about, darting from perch to perch.

Scalyhead Sculpin Female

Brownish color variation.

DISTINCTIVE FEATURES: 1. Orangish area below lateral line. 2. Broad band of scales between dorsal fin and lateral line. 3. Snout somewhat elongated compared to similar appearing sculpins. 4. Cirri on head not large or conspicuous.

DESCRIPTION: Mottled, spotted and blotched earthtone shades, often with large patches of red to lavender, and several dark saddle markings on back to lateral line; pale gray spots below. Can change color and markings to blend with background. No scales on head. Embedded scales along lateral line have thread-like extensions.

ABUNDANCE & DISTRIBUTION: Common California; occasional to uncommon north to Washington. Also south to central Baja.

HABITAT & BEHAVIOR: Inhabit rocky reefs, outcroppings and coastlines. Unlike most sculpins, actively move about, darting from perch to perch.

REACTION TO DIVERS: Wary; dart to new perch when approached. Stalking with slow nonthreatening movements may allow a closer view.

Bulbous, Spiny-Headed Bottom-Dwellers

SCALYHEAD SCULPIN
Artedius harringtoni
FAMILY:
Sculpin – Cottidae

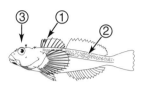

SIZE: 2-3 in., max. 4 in.
DEPTH: 0-70 ft.

Scalyhead Sculpin Male
Note large cirri on head.

CORALLINE SCULPIN
Artedius corallinus
FAMILY:
Sculpin – Cottidae

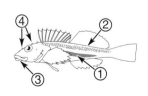

SIZE: 2-4 in., max. 5 1/2 in.
DEPTH: 0-70 ft.

Sculpin

DISTINCTIVE FEATURES: 1. Blunt snout. 2. Broad band of scales between dorsal fin and lateral line. 3. Obvious scales from top of head to foredorsal fin.

DESCRIPTION: Mottled, spotted and blotched earthtone shades, often with large patches of red to lavender or orange; several dark saddle markings on back to lateral line, usually small pale spots below lateral line. Can change color and markings to blend with background. Embedded scales along lateral line have thread-like extensions.

ABUNDANCE & DISTRIBUTION: Occasional central and southern California. Also south to North Baja.

HABITAT & BEHAVIOR: Inhabit rocky reefs, outcroppings and coastlines. Unlike most sculpins, actively dart from perch to perch.

REACTION TO DIVERS: Wary; dart to new perch when approached. Stalking with slow nonthreatening movements may allow a closer view.

Snubnose Sculpin
Red color variation.

DISTINCTIVE FEATURES: Slender, tapered elongated body. **1. Lower spines of pectoral fins long, extending well beyond fin membrane. 2. Slanting folds of flesh along lower sides of body give washboard texture. 3. Spinous dorsal fin rounded and separated from long, straight-edged soft dorsal.**

DESCRIPTION: Mottled and blotched shades of gray to olive or brown, often with five saddles on back; whitish underside.

ABUNDANCE & DISTRIBUTION: Uncommon Bering Sea, Alaska, to Washington.

HABITAT & BEHAVIOR: Inhabit sand, silt and mud bottoms. Rest on bottom, blending with surroundings, often propped up on pectoral fins and tail. Rarely in open during day; nocturnally active.

REACTION TO DIVERS: Remain still, apparently relying on camouflage. Bolt only when closely approached or molested.

Bulbous, Spiny-Headed Bottom-Dwellers

SNUBNOSE SCULPIN
Orthonopias triacis
FAMILY:
Sculpin – Cottidae

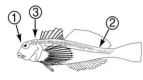

SIZE: 2-3 in., max. 4 in.
DEPTH: 0-100 ft.

Snubnose Sculpin
Brown color variation.

ROUGHSPINE SCULPIN
Triglops macellus
FAMILY:
Sculpin – Cottidae

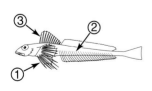

SIZE: 3-6 in., max. 8 in.
DEPTH: 60-300 ft.

Sculpin

DISTINCTIVE FEATURES: 1. Rough scales cover upper head and back. 2. First dorsal fin spine long. 3. V-shaped indention between third and fourth dorsal fin spines, and another notch separating spinous and soft parts.

DESCRIPTION: Mottled and blotched shades of gray to greenish gray and brown often with saddle markings on back, pale sides and belly. Can change color and markings to blend with background. Branched upper spine on cheek. Breeding males often develop a red to orange, lavender or white area above eyes; they may also have irregular red to orange or lavender bands or markings on back and sides.

ABUNDANCE & DISTRIBUTION: Abundant to common northern British Columbia, Canada, to southern California. Also to southern Baja.

HABITAT & BEHAVIOR: Inhabit sand, silt and mud bottoms, usually below 25 feet. More commonly in open at night.

REACTION TO DIVERS: Remain still, apparently relying on camouflage. Bolt only when closely approached or molested.

Roughback Sculpin Breeding Male
Note orange patch between eyes.

DISTINCTIVE FEATURES: 1. All fins (except anal) have bars formed by aligned spots on fin rays. 2. Foredorsal fin has nearly smooth outline with a dark patch on upper forefin and another at upper rear. 3. Double row of large scales run from under foredorsal fin to tail.

DESCRIPTION: Mottled and blotched shades of gray to brown and white, often with saddle markings on back; usually small pale spots below lateral line. Can change color and markings to blend with background. Embedded scales along lateral line have thread-like extensions.

ABUNDANCE & DISTRIBUTION: Uncommon northern Washington to Bering Sea.

HABITAT & BEHAVIOR: Inhabit sand plains, silt, gravel or crushed shell bottoms. Rest motionlessly on bottom blending with substrate. Rarely in open during day; nocturnally active.

REACTION TO DIVERS: Remain still, apparently relying on camouflage. Bolt only when closely approached or molested.

Bulbous, Spiny-Headed Bottom-Dwellers

ROUGHBACK SCULPIN
Chitonotus pugetensis
FAMILY:
Sculpin – Cottidae

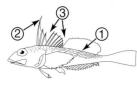

SIZE: 4-7 in., max. 9 in.
DEPTH: 0-450 ft.

Roughback Sculpin Breeding Male
Note pinkish patch between eyes.

NORTHERN SCULPIN
Icelinus borealis
FAMILY:
Sculpin – Cottidae

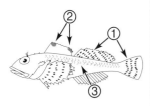

SIZE: 2-3 in., max. 4 in.
DEPTH: 25-800 ft.

67

Sculpin

DISTINCTIVE FEATURES: 1. First two spines of dorsal fin extremely long, first usually longest. 2. Head depressed behind eyes, with two paddle-like cirri behind eyes and two or more prominent spines. (Similar Threadfin Sculpin [next] may have cirri or bumps, but no spines behind eyes.) **MALE: 3. Dusky to dark spot on mid-foredorsal fin.**

DESCRIPTION: Mottled and blotched shades of gray to brown, occasionally have patches of brilliant red, lavender or rose, often several saddles on back; usually pale spots below lateral line. Base of pectoral fin silvery, outer half flecked with yellow or gold. Slender, tapered elongated body.

ABUNDANCE & DISTRIBUTION: Occasional California; uncommon north to central British Columbia, Canada. Also south to central Baja.

HABITAT & BEHAVIOR: At night, rest in open on deep sandy and silty bottoms. Occasionally found during day camouflaged on rocky reefs.

REACTION TO DIVERS: Apparently mesmerized by light, they tend to remain still until closely approached before darting.

DISTINCTIVE FEATURES: 1. First two spines of dorsal fin extremely long, often of nearly equal length. 2. Upper head behind eyes not concave and without spines. (Similar Spotfin Sculpin [previous] distinguished by depressed head behind eyes and with two or more prominent spines.)

DESCRIPTION: Mottled and blotched shades of gray to brown, often several saddles on back; usually pale and unmarked below lateral line.

ABUNDANCE & DISTRIBUTION: Rare (more common below safe diving limits) southern California north to Vancouver Island, Canada.

HABITAT & BEHAVIOR: Inhabit deep sand and mud bottoms. Rest on belly on bottom or propped up on pectoral fins. Nocturnal, rarely observed during the day.

REACTION TO DIVERS: Apparently mesmerized by light, they tend to remain still until closely approached before darting away.

Bulbous, Spiny-Headed Bottom-Dwellers

SPOTFIN SCULPIN
Icelinus tenuis
FAMILY:
Sculpin – Cottidae

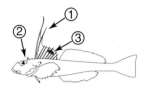

SIZE: 2-4 in., max. 5 1/2 in.
DEPTH: 40-1,200 ft.

Spotfin Sculpin Male [right]
Note dusky spot on mid-foredorsal fin.

Spotfin Sculpin Color Variations [left and right]
Note differing lengths of elongated first two dorsal fin spines.

THREADFIN SCULPIN
Icelinus filamentosus
FAMILY:
Sculpin – Cottidae

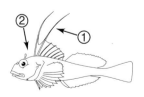

SIZE: 3-7 in., max. 10 3/4 in.
DEPTH: 60-1,200 ft.

Sculpin

DISTINCTIVE FEATURES: Long, slender tapering body. **1. Three pale bands below eye. 2. Slanting rows of fused, serrated scales along lower sides of body give washboard texture. 3. Long spinous dorsal fin virtually equal in length to soft dorsal fin.**
DESCRIPTION: Shades of yellow to greenish yellow, greenish brown, orange, reddish brown and brown; occasionally marked with red; usually with six to eight saddles on back; often yellow to orange tail.
ABUNDANCE & DISTRIBUTION: Common to occasional southeast Alaska to central California.
HABITAT & BEHAVIOR: Inhabit wide range of habitats from tide pools to rocky slopes, kelp beds, sheer rock faces and ledge overhangs. Especially common in tidal channels. Dart about bottom, occasionally stopping to rest on pectoral fins; or may hang vertically on wall faces or upside-down under overhangs.
REACTION TO DIVERS: Wary; dart away when approached. Occasionally stalking with slow nonthreatening movements will allow a close view.

Longfin Sculpin
Brown color variation.

DISTINCTIVE FEATURES: 1. Numerous cirri on top of head. 2. Short pointed spines and cirri between dorsal fin and lateral line.
DESCRIPTION: Gray to grayish green to greenish brown and spotted and lightly blotched with white, occasionally tinted pink or yellow.
ABUNDANCE & DISTRIBUTION: Common central and southern California; uncommon northern California. Also south to central Baja.
HABITAT & BEHAVIOR: Inhabit tide pools and shallow rocky reefs and kelp beds. Frequently nestled in leafy algae growth. Flit about, often briefly stopping to perch on hard surfaces.
REACTION TO DIVERS: Wary; bolt and resettle nearby when closely approached. Stalking with slow nonthreatening movements may allow close view.
SIMILAR SPECIES: Mosshead Sculpin, *C. globiceps*, distinguished by numerous cirri along fore lateral line. Southern California to Alaska, but rarely south of San Francisco. Bald Sculpin, *C. recalvus*, distinguished by lack of short spines or cirri between dorsal fins and lateral line and cirri only on rear half of head. Southern Oregon to central Baja, but rarely north of Santa Cruz. Both species are primarily tide pool dwellers.

Bulbous, Spiny-Headed Bottom-Dwellers

LONGFIN SCULPIN
Jordania zonope
FAMILY:
Sculpin – Cottidae

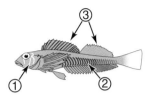

SIZE: 2-5 in., max. 6 in.
DEPTH: 0-130 ft.

Longfin Sculpin
Yellow color variation.

WOOLLY SCULPIN
Clinocottus analis
FAMILY:
Sculpin – Cottidae

SIZE: 2-4 in., max. 7 in.
DEPTH: 0-65 ft.

71

Sculpin

DISTINCTIVE FEATURES: Bulbous head and tapered body. Scaleless. **1. Low spinous dorsal fin continuous with somewhat taller soft portion.**
DESCRIPTION: Shades of brown to gray and occasionally tints of pink; fins pale and lightly spotted.
ABUNDANCE & DISTRIBUTION: Uncommon to rare Bering Sea, Alaska, to Puget Sound, Washington.
HABITAT & BEHAVIOR: Inhabit wide range of habitats from rocky to soft bottoms. Nocturnally active, swim slowly above bottom. At night attracted to light.
REACTION TO DIVERS: Not afraid; at night attracted to divers' handlights.
NOTE: Formerly classified in the genus *Gilbertidia*.

DISTINCTIVE FEATURES: Slender, tapered elongated body. **1. Pectoral fins join on underside. 2. Generally seven whitish blotches on back and several smaller spots below lateral line. 3. Pointed snout.**
DESCRIPTION: Olive to olive-brown, brown, golden-brown or other earthtone shades; change color to blend with surroundings.
ABUNDANCE & DISTRIBUTION: Occasional southeast Alaska to southern California. Can be locally common in kelp beds.
HABITAT & BEHAVIOR: Inhabit areas of kelp, sea lettuce and other leafy algae; also in tide pools, floats of seaweed and on dock pilings. Rest on leaves, changing color to blend almost perfectly with background. Occasionally dart from leaf to leaf.
REACTION TO DIVERS: Wary; dart away only when closely approached. Occasionally stalking with slow nonthreatening movements will allow a close view.

DISTINCTIVE FEATURES: 1. First two extremely long dorsal fin spines form a spike-like projection. 2. Red to blue spots on spines of dorsal fins aligned to form diagonal bands.
DESCRIPTION: Lightly mottled and spotted in shades of olive to brown or orange-brown over cream undercolor, with brown to brownish red saddles or bars across body. Slender, tapered elongated body.
ABUNDANCE & DISTRIBUTION: Occasional southern California. Also south to northern Baja.
HABITAT & BEHAVIOR: Inhabit shallow sand flats and eelgrass beds; also rocky reefs, especially around kelp beds.
REACTION TO DIVERS: Very shy; scurry away when approached. On rare occasions, stalking with slow nonthreatening movements may allow close view.

Bulbous, Spiny-Headed Bottom-Dwellers

SOFT SCULPIN
Psychrolutes sigalutes
FAMILY:
Sculpin – Cottidae

SIZE: 1-2 in.,
max. 3¼ in.
DEPTH: 3-740 ft.

MANACLED SCULPIN
Synchirus gilli
FAMILY:
Sculpin – Cottidae

SIZE: 1-2½ in.,
max. 3 in.
DEPTH: 0-80 ft.

LAVENDER SCULPIN
Leiocottus hirundo
FAMILY:
Sculpin – Cottidae

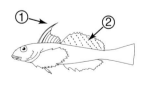

SIZE: 4-7 in.,
max. 10 in.
DEPTH: 0-120 ft.

Sculpin

DISTINCTIVE FEATURES: 1. Foredorsal fin extremely long. 2. Long, ruffled second dorsal fin. 3. Dark band runs across eye and onto cheek.
DESCRIPTION: Lightly mottled shades of brown to gray or cream on back gradating to paler shades on sides and belly. Slender, tapered elongated body.
ABUNDANCE & DISTRIBUTION: Common Bering Sea, Alaska, to Puget Sound; occasional to uncommon to southern California.
HABITAT & BEHAVIOR: Inhabit shallow rocky bottoms, especially near base of tall rocky outcroppings; occasionally around dock pilings and jetties. Nocturnally active, swim with graceful undulations of soft dorsal fin while keeping foredorsal erect. Hide during day in caves, deep crevices and other protective recesses.
REACTION TO DIVERS: Not shy; can be closely approached at night with slow nonthreatening movements. Bolt only if disturbed.

DISTINCTIVE FEATURES: 1. Several long, straight, hair-like cirri on snout and chin. 2. Tall spinous dorsal fin with deep notch before large, rounded soft dorsal fin. 3. A row of several silvery to white spots on side behind pectoral fin.
DESCRIPTION: Greenish golden-brown to brown back, occasionally marked with dark patches, changing to a coppery-brown to yellowish underside. Large anal fin matches length of soft dorsal fin.
ABUNDANCE & DISTRIBUTION: Common northern California to Bering Sea, Alaska. Also to Sea of Japan.
HABITAT & BEHAVIOR: Inhabit shallow rocky bottoms in areas of luxuriant, leafy algae growth. Nestle down in algae blades, camouflaging with background.
REACTION TO DIVERS: Remain still, apparently relying on camouflage. Move only when closely approached or molested.

DISTINCTIVE FEATURES: Short, stocky, curled body. **1. Large, bulbous head with tapered snout.**
DESCRIPTION: Mottled and streaked in shades of cream to orange-brown to dark brown; bright orange bar on base of tail. Pectoral fin rays without connecting membrane.
ABUNDANCE & DISTRIBUTION: Occasional Alaska to Puget Sound, Washington; rare south to southern California. Also Japan.
HABITAT & BEHAVIOR: Inhabit rocky bottoms and areas of sand mixed with rubble. "Crawl" about bottom on pectoral fin rays. Often take shelter in empty shell casings, especially those of the Giant Barnacle, *Balanus nubilus*. Prefer waters in low fifties or below and are consequently rarely observed within safe diving limits south of Puget Sound.
REACTION TO DIVERS: Appear unafraid; continue normal activities, scuttling away to shelter only if molested.

Bulbous, Spiny-Headed Bottom-Dwellers

SAILFIN SCULPIN
Nautichthys oculofasciatus
FAMILY:
Sculpin – Cottidae

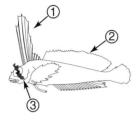

SIZE: 2-6 in., max. 8 in.
DEPTH: 3-360 ft.

SILVERSPOTTED SCULPIN
Blepsias cirrhosus
FAMILY:
Sculpin – Cottidae

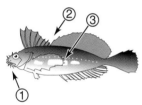

SIZE: 3-6 in., max. 8 in.
DEPTH: 0-125 ft.

GRUNT SCULPIN
Rhamphocottus richardsoni
FAMILY:
Sculpin – Cottidae

SIZE: 2-3 in., max. 3 1/2 in.
DEPTH: 0-540 ft.

IDENTIFICATION GROUP 3

Eels and Eel-Like Bottom-Dwellers
Prickleback – Gunnel – Others

This ID Group consists of bottom-dwelling fish with long, compressed bodies and continuous, or almost continuous, dorsal, tail and anal fins.

FAMILY: Prickleback — Stichaeidae
8 Species Included

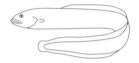

Pickleback (typical shape)

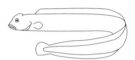

Slender Cockscomb

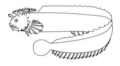

Mosshead Warbonnet

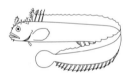

Decorated Warbonnet

    In addition to the pricklebacks, this family also includes warbonnets and cockscombs which can be distinguished by their fleshy head appendages. The majority of these bottom-dwelling fish live in the cold waters of the North Pacific. Long continuous dorsal fins supported by sharp, rigid spines run the length of their long compressed bodies. This characteristic distinguishes family members from similar gunnels that have flexible spines. Both the long anal fin, which extends at least half the body length, and dorsal fin end just before the rounded tail. These fish vary from a few inches to slightly over one and a half feet in length.
    They take shelter under rocks, in recesses, or mix in with bottom debris or marine plant growth in a variety of habitats. Their bodies occasionally curl as they move about the bottom. Most species can easily be identified by markings and head appendages.

FAMILY: Gunnel — Pholidae
6 Species Included

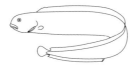

Gunnel (typical shape)

Gunnels are a small family of shallow, cold-water, bottom-dwelling fish. Most species inhabit North Pacific waters. They have compressed elongated bodies with a single dorsal fin beginning just behind the head and extending to the tail. This fin is supported by flexible spines that distinguish them from the similar appearing pricklebacks that have sharp, rigid, spike-like spines. Dorsal and anal fins join the nearly circular tail. Gunnels range in length from a few inches to about one and a half feet.

These slender bottom-dwelling fish often take shelter under rocks, in recesses, or mix with bottom debris or marine plant growth in a variety of habitats. Their bodies curl in a snake-like fashion as they move about the bottom. Most are colorful and display distinctive markings, making identification relatively simple.

FAMILY: Others
10 Species Included

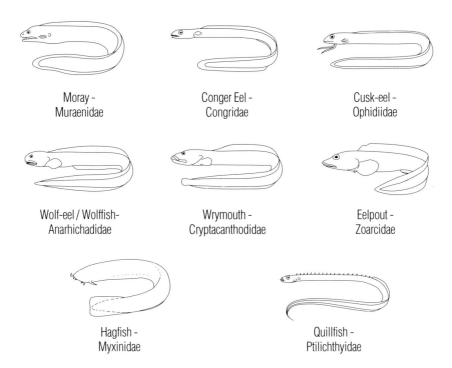

Moray - Muraenidae

Conger Eel - Congridae

Cusk-eel - Ophidiidae

Wolf-eel / Wolffish - Anarhichadidae

Wrymouth - Cryptacanthodidae

Eelpout - Zoarcidae

Hagfish - Myxinidae

Quillfish - Ptilichthyidae

Moray – Conger Eel – Cusk-Eel

DISTINCTIVE FEATURES: Light to dark brown to green **1. No pectoral fins.**
DESCRIPTION: Often mottled. Mouth filled with numerous sharp teeth.
ABUNDANCE & DISTRIBUTION: Abundant Santa Catalina and San Clemente Islands, California; common southern California. Also south to southern Baja.
HABITAT & BEHAVIOR: Inhabit rocky reefs. Lurk in caves, crevices and other protective recesses. Often rest on bottom with head extending from protective recess. Constantly open and close mouth, an action required for respiration — not a threat.
REACTION TO DIVERS: Curious; often peer out from hole or crack with head and forebody exposed, retreat only if very closely approached. May bite if molested or speared.

DISTINCTIVE FEATURES: Reddish brown to grayish brown. **1. Have pectoral fins.**
DESCRIPTION: Rear of pectoral fins extend to just below beginning of dorsal fin. Rounded tail.
ABUNDANCE & DISTRIBUTION: Occasional southern California. Also south to Baja and Gulf of California.
HABITAT & BEHAVIOR: Inhabit sand and soft mud bottoms. Can burrow tailfirst into sand or mud.
REACTION TO DIVERS: Generally appear unafraid; burrow into soft bottom when closely approached.

DISTINCTIVE FEATURES: 1. Dark brown spots cover body. 2. Ventral fins have evolved into two long, whisker-like appendages.
DESCRIPTION: Cream to nearly translucent undercolor. Dorsal, anal and pointed tail fin are continuous; pectoral fin far forward. (Similar appearing species have pelvic fins further back on body.)
ABUNDANCE & DISTRIBUTION: Occasional northern Oregon to southern California. Also south to central Baja.
HABITAT & BEHAVIOR: Inhabit mud, clay, sand and sand/rubble bottoms. Live in mucus-lined burrows which they enter tailfirst. If alarmed can bury directly into sand tailfirst. Forage about in open at night and occasionally on overcast days.
REACTION TO DIVERS: Shy; usually move away from light and occasionally bury. In dim light they appear less concerned with diver's presence if no handlight is used.

Eels & Eel-like Bottom-Dwellers

CALIFORNIA MORAY
Gymnothorax mordax
FAMILY:
Moray - Muraenidae

SIZE: 2 - 4 ft., max. 5 ft.
DEPTH: 2 - 130 ft.

CATALINA CONGER
Gnathophis catalinensis
FAMILY:
Conger Eel – Congridae

SIZE: 8 - 14 in., max. 16 1/2 in.
DEPTH: 30 - 1,200 ft.

SPOTTED CUSK-EEL
Chilara taylori
FAMILY:
Cusk-Eel – Ophidiidae

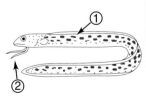

SIZE: 4 - 10 in., max. 14 in.
DEPTH: 3 - 1,000 ft.

Wolf-eel – Wrymouth

DISTINCTIVE FEATURES: 1. Large, bulbous head with large mouth and numerous blunt teeth. 2. Dark centered ocellated spots on body and fins.

DESCRIPTION: Usually shades of gray, occasionally brown; larger individuals often somewhat mottled. No ventral fins. Virtually continuous dorsal, tail and anal fin only slightly notched before tiny tail. **MALE:** Head whitish, flabby and lumpy compared to darker and more lean head of female. **JUVENILE:** Orange to orangish brown with dark spots.

ABUNDANCE & DISTRIBUTION: Common Aleutian Islands, Alaska, to southern California. Also to Sea of Japan.

HABITAT & BEHAVIOR: Inhabit dens in crevices, caves and other recesses in rocky, boulder-strewn areas. Mating couples occupy the same den, apparently staying together for life. Females lay large egg masses in den which couple protects until hatching.

REACTION TO DIVERS: Generally appear unafraid, but will retreat into den when closely approached or molested. Can be "trained" to be hand-fed by divers.

Wolf-eel Juvenile
Note reddish color.

DISTINCTIVE FEATURES: 1. Large, wide, flattened head with projecting lower jaw and upturned mouth that extends well past eye.

DESCRIPTION: Shades of brown to gray with rows of small dark blotches along sides; occasionally tinted or marked with violet and/or yellow. No ventral fins. Virtually continuous dorsal, tail and anal fin; tail is defined by notches and nearly circular shape; long anal fin is more than half the length of dorsal.

ABUNDANCE & DISTRIBUTION: Occasional Aleutian Islands, Alaska, to northern California.

HABITAT & BEHAVIOR: Inhabit compact silt and mud bottoms, usually below 40 feet. Live in multi-tunnelled burrows which may have several entrances. Occasionally rest with only head protruding from burrow entrance. Rarely leave burrow.

REACTION TO DIVERS: Lethargic; usually do not move if approached with slow nonthreatening movements. Rapidly withdraw deep into burrow when threatened.

NOTE: Formerly classified in the genus *Debleps*.

Eels & Eel-like Bottom-Dwellers

WOLF-EEL
Anarrhichthys ocellatus
FAMILY:
Wolffish – Anarhichadidae

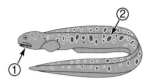

SIZE: 2½ - 5 ft.,
max. 8 ft.
DEPTH: 0 - 700 ft.

Wolf-eel
Couple in den guarding eggs. Note male [right] distinguished by whitish, flabby and lumpy head.

GIANT WRYMOUTH
Cryptacanthodes giganteus
FAMILY:
Wrymouth –
Cryptacanthodidae

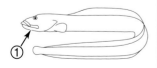

SIZE: 2 - 3 ft.,
max. 3¾ ft.
DEPTH: 20 - 420 ft.

Prickleback

DISTINCTIVE FEATURES: 1. Uneven ridge runs from snout to rear of head.
DESCRIPTION: Highly variable colors from nearly solid shades of olive to gray and black to blotched and barred, often with two dark bars below eye and occasionally with orange spots on back and edging on fins. No ventral fins. Young may have white stripe running up snout and head and along border of dorsal fin.
ABUNDANCE & DISTRIBUTION: Common to occasional southern Oregon to central California; rare southern California. Also south to central Baja.
HABITAT & BEHAVIOR: Inhabit inshore areas from tide pools to shallow rocky reefs. Remain in, or near same crack, crevice, cave or under rock for years.
REACTION TO DIVERS: Shy; retreat to cover when approached. On rare occasions, apparently when relying on camouflage, can be approached with slow nonthreatening movements as they peer out from shelter.

DISTINCTIVE FEATURES: 1. Two dark bars with white to cream outline radiate from behind and below each eye. 2. White bar on base of tail.
DESCRIPTION: Shades of brown. Tiny pectoral fins; dorsal and anal fins continuous with rounded tail.
ABUNDANCE & DISTRIBUTION: Occasional southern Alaska to southern California. Also south to northern Baja.
HABITAT & BEHAVIOR: Inhabit rocky substrates mixed with sea grass and algae growth. Most common intertidal to 20-foot depths. Often in cracks, crevices, recesses and under rocks, or lurk in tangles of lush plant growth.
REACTION TO DIVERS: Very shy; when approached, rapidly move away and wiggle under rocks or other cover.
SIMILAR SPECIES: Rock Prickleback, *X. mucosus*, distinguished by two bars with dark outlines radiating from eyes. Greenish gray to brown; grow to two feet. Inhabit depths to 65 feet. Ribbon Prickleback, *Phytichthys chirus*, distinguished by several light and dark lines radiating from behind and below eyes. Olive green to brown; grow to eight inches. Inhabit depths to 45 feet. Rare.

Eels & Eel-like Bottom-Dwellers

MONKEYFACE PRICKLEBACK
Cebidichthys violaceus
FAMILY:
Prickleback – Stichaeidae

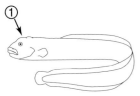

SIZE: 1 - 2 ft.,
max. 2½ ft.
DEPTH: 0 - 80 ft.

Monkeyface Prickleback Juvenile
Note white stripe on head and dorsal fin. [right]

Note orange spots on back. [far left]

Large dark adult, virtually without markings. [near left]

BLACK PRICKLEBACK
Xiphister atropurpureus
FAMILY:
Prickleback – Stichaeidae

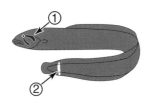

SIZE: 4 - 8 in.,
max. 1 ft.
DEPTH: 0 - 35 ft.

Cockscomb – Warbonnet

DISTINCTIVE FEATURES: 1. Fleshy crest from snout to top of head. 2. Two darkish bars extend below each eye.
DESCRIPTION: Shades of black to gray or brown, often mottled and occasionally with purple overtones; dorsal fin usually paler shade joining pale saddles on back. Change color and markings to blend with background. Breeding males often have orange to orangish fins; females have white spots [pictured].
ABUNDANCE & DISTRIBUTION: Occasional Alaska to southern California.
HABITAT & BEHAVIOR: Most common on rocky bottoms from intertidal to ten-foot depths; occasionally rocky reefs to 100 feet. Live in cracks, crevices, recesses, under rocks and inside discarded bottles.
REACTION TO DIVERS: Very shy; when approached, rapidly move away and wiggle under rocks or other cover.

DISTINCTIVE FEATURES: 1. Fleshy crest from snout to top of head. 2. Two or three dark bars on foredorsal fin. 3. Pale bands on lower lip and under jaw.
DESCRIPTION: Variable shades from dark gray to brown and occasionally orange to red. Often have pale and dark bars on cheek and pale bars on back and dorsal fin. Change color and markings to blend with background.
ABUNDANCE & DISTRIBUTION: Occasional Alaska to northern California.
HABITAT & BEHAVIOR: Inhabit rocky substrates, often in areas with strong currents. Quite secretive; live in cracks, crevices, recesses, and under rocks. Uncommon intertidally or in shallow water.
REACTION TO DIVERS: Very shy; when approached, rapidly move away and wiggle under rocks or other cover.

DISTINCTIVE FEATURES: 1. Numerous cirri on head to dorsal fin. 2. Row of pale to white, darkly outlined bars on lower sides. MALE : 3. About twelve dark, goldish ringed spots on dorsal fin [pictured]. FEMALE: 4. About twelve dark bars on dorsal fin.
DESCRIPTION: Mottled shades of brown. Dark bar below eye, and may have several more on gill cover. Have ventral fins. Dorsal and anal fin separated from rounded tail by shallow notch; long anal fin is about three-fourths the length of dorsal.
ABUNDANCE & DISTRIBUTION: Occasional Aleutian Islands, Alaska, to southern California.
HABITAT & BEHAVIOR: Inhabit shallow areas with numerous small protective recesses, including debris under docks. Often perch in small opening with only their heads extended; favorite haunts include tube worm holes, empty shells, small crevices, bottles and cans.
REACTION TO DIVERS: Shy; dart deep into recess when closely approached. May allow close view with slow nonthreatening movements, but, once frightened, rarely reappear from hiding.

Eels & Eel-like Bottom-Dwellers

HIGH COCKSCOMB
Anoplarchus purpurescens
FAMILY:
Prickleback – Stichaeidae

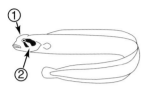

SIZE: 3-5 in.,
max. 7 ¾ in.
DEPTH: 0-100 ft.

SLENDER COCKSCOMB
Anoplarchus insignis
FAMILY:
Prickleback – Stichaeidae

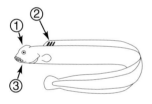

SIZE: 2½ - 4 in.,
max. 5 in.
DEPTH: 0-100 ft.

MOSSHEAD WARBONNET
Chirolophis nugator
FAMILY:
Prickleback – Stichaeidae

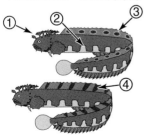

SIZE: 2-4 in., max. 6 in.
DEPTH: 0-260 ft.

Warbonnet – Prickleback

DISTINCTIVE FEATURES: 1. Large bushy cirri between and in front of eyes. 2. Bushy, fleshy projections (also known as cirri) on first four dorsal fin spines.
DESCRIPTION: Mottled or barred in shades of brown. Dark band extends below eye; dorsal fin has large dark bars; fan-like pectoral fins have design of dark, expanding, semi-circular bands. Numerous small cirri on top of head. Have ventral fins. Dorsal and anal fin separated from rounded tail by shallow notch; long anal fin is about three-fourths the length of dorsal.
ABUNDANCE & DISTRIBUTION: Occasional Aleutian Islands, Alaska, to northern California. Also to Siberia, Russia.
HABITAT & BEHAVIOR: Inhabit rough, rocky areas with numerous crevices, caves and recesses. Perch in entrances of protective recesses, including the openings of large sponges.
REACTION TO DIVERS: Shy; dart deep into recess when closely approached. May allow close view with slow nonthreatening movements, but once frightened seldom reappear.

Decorated Warbonnet
Detail of head cirri.
Color variation.

DISTINCTIVE FEATURES: 1. Broken stripe of dark, dash-like markings along midbody. 2. Four to five narrow, irregular bars on tail.
DESCRIPTION: Shades of gray, occasionally brownish to light green; back darker, with darkish blotches; pale lower body and belly. Have pectoral fins. Dorsal and anal fin are separated from wedge-shaped tail; long anal fin is about three-fourths the length of dorsal.
ABUNDANCE & DISTRIBUTION: Abundant (summer and early fall) Aleutian Islands, Alaska, to northern California.
HABITAT & BEHAVIOR: Inhabit soft bottoms of sand, silt or mud, especially shallow bays and inlets in summer and early fall; migrate to deeper water winter and spring. Perch on bottom in stiff "tripod stance" on pectoral, ventral and tail fins.
REACTION TO DIVERS: Quite wary; dart away when closely approached. At night, apparently mesmerized by handlight, remain still and allow close view.
NOTE: Also commonly called "Pacific Snake Prikleback."

Eels & Eel-like Bottom-Dwellers

DECORATED WARBONNET
Chirolophis decoratus
FAMILY:
Pickleback – Stichaeidae

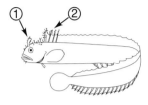

SIZE: 4-10 in., max. 16 in.
DEPTH: 5-300 ft.

Decorated Warbonnet
Color variation.

SNAKE PRICKLEBACK
Lumpenus sagitta
FAMILY:
Pickleback – Stichaeidae

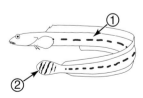

SIZE: 6-10 in.,
max. 20 in.
DEPTH: 3-680 ft.

87

Eelpout – Gunnel

DISTINCTIVE FEATURES: 1. Black spot on foredorsal fin, and remainder of fin edged in black. 2. Black belly.
DESCRIPTION: Shades of gray to brown. Large, rounded, fan-like pectoral fin. Continuous dorsal, tail and anal fin.
ABUNDANCE & DISTRIBUTION: Common Gulf of Alaska to northern Baja.
HABITAT & BEHAVIOR: Inhabit sand, mud, gravel and rubble-strewn bottoms. Nocturnal, move about bottom in search of prey. May lie on bottom curving body in a snake-like manner.
REACTION TO DIVERS: Wary; but occasionally allow a slow nonthreatening approach.
SIMILAR SPECIES: Black Eelpout, *Lycodes diapterus,* distinguished by pale, wavy bar markings on back, and shallow V-shaped notch in pectoral fin; Bering Sea to southern California. Shortfin Eelpout, *L. brevipes,* distinguished by large dark dot on foredorsal fin, and pale bars across dorsal fin and back; Bering Sea to Oregon. Both inhabit deep sand and mud bottoms; rare within safe diving limits.

DISTINCTIVE FEATURES: Solid shades of yellow-green to green or reddish brown. **1. Tiny pectoral fin.**
DESCRIPTION: Often short dark marking below eye and occasionally a row of spots along side. No ventral fins. Continuous dorsal, tail and anal fin; tail is defined by slightly longer rays; anal fin slightly less than half the length of dorsal.
ABUNDANCE & DISTRIBUTION: Common to occasional British Columbia, Canada, to southern California. Also south to central Baja.
HABITAT & BEHAVIOR: Take refuge in the lush vegetation of rockweed, sea lettuce and leafy red algae beds in shallow inshore areas, including tide pools. Change color to blend with algae.
REACTION TO DIVERS: Very shy; rapidly retreat to cover when approached.
SIMILAR SPECIES: Kelp Gunnel, *Ulvicola sanctaerosae,* distinguished by lack of pectoral and ventral fins.
NOTE: Formerly classified in the genus *Xererpes.*

DISTINCTIVE FEATURES: 1. Usually a dark bar below each eye. 2. Generally a row of dark and/or pale spots along midbody.
DESCRIPTION: Vary in shades of green or maroon or tobacco brown. No ventral fins. First spine of anal fin large and grooved like a fountain pen point. Continuous dorsal, tail and anal fin; tail is defined by slightly longer rays; anal fin about half the length of dorsal.
ABUNDANCE & DISTRIBUTION: Common Gulf of Alaska to southern California.
HABITAT & BEHAVIOR: Take on color of vegetation they inhabit: eelgrass and sea lettuce beds (usually shades of green); leafy, red algae beds (usually shades of maroon); stands of kelp (usually shades of tobacco brown). When sea grasses/algae are not present (often in winter), inhabit rocky areas, lurking under rocks and in protective recesses.
REACTION TO DIVERS: Remain still, apparently relying on camouflage. Slow nonthreatening approach usually allows close observation.

Eels & Eel-like Bottom-Dwellers

BLACKBELLY EELPOUT
Lycodopsis pacifica
FAMILY:
Eelpout – Zoarcidae

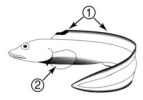

SIZE: 8-15 in.,
max. 1½ ft.
DEPTH: 40-1,320 ft.

ROCKWEED GUNNEL
Apsdichthys fucorum
FAMILY:
Gunnel – Pholidae

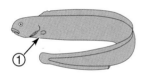

SIZE: 3-6 in., max. 9 in.
DEPTH: 0-30 ft.

PENPOINT GUNNEL
Apodichthys flavidus
FAMILY:
Gunnel – Pholidae

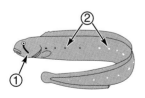

SIZE: 4-8 in.,
max. 1½ ft.
DEPTH: 0-60 ft.

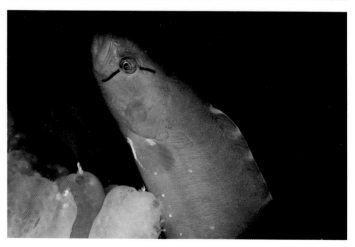

89

Gunnel

DISTINCTIVE FEATURES: 1. Numerous pale, narrow saddles across dorsal fin and back with dark specks on back.
DESCRIPTION: Shades of brilliant red to maroon, orange, orangish brown or reddish brown, often with some silver markings; belly pale. Small ventral fins. Continuous dorsal, tail and anal fin; tail is defined by slightly longer rays; anal fin over half the length of dorsal.
ABUNDANCE & DISTRIBUTION: Occasional to uncommon southeastern Alaska to northern California.
HABITAT & BEHAVIOR: Inhabit rocky areas with leafy, red algae. Often mixed in with algae, occasionally in open resting on rocky surface.
REACTION TO DIVERS: Wary; retreat into tangles of algae or protective recess when approached. Occasionally curious, peering out from protective cover where they can be closely approached with slow nonthreatening movements.

Crescent Gunnel
Orange-brown color variation.

DISTINCTIVE FEATURES: 1. Series of prominent, roundish pale blotches with black outlines along upper back and base of dorsal fin. 2. Pale bars to large spots along mid-sides. 3. Dark bar below each eye.
DESCRIPTION: Shades of yellow-green to yellow-brown, orange-brown or brown, with row of pale bars or spots along midbody. Tiny ventral fins; pectorals relatively large. Continuous dorsal, tail and anal fin; tail is defined by slightly longer rays; anal fin about half the length of dorsal.
ABUNDANCE & DISTRIBUTION: Aleutian Islands southward to northern California.
HABITAT & BEHAVIOR: Inhabit eelgrass beds or rocky areas of leafy algae; also around jetties and under docks, living in jars, cans, tires and other debris. May mix in with eelgrass or leafy algae, occasionally hide under rocks or in protective recesses. More commonly in open at night.
REACTION TO DIVERS: Wary; retreat into tangles of algae or protective recess when approached. Occasionally curious, peering out from protective cover where they can be closely approached with slow nonthreatening movements.

Eels & Eel-like Bottom-Dwellers

LONGFIN GUNNEL
Pholis clemensi
FAMILY:
Gunnel – Pholidae

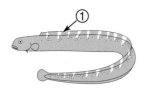

SIZE: 3-5 in., max. 5 in.
DEPTH: 25 - 200 ft.

Longfin Gunnel
Orange color variation.

CRESCENT GUNNEL
Pholis laeta
FAMILY:
Gunnel – Pholidae

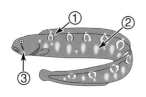

SIZE: 3-5 in., max. 10 in.
DEPTH: 0 - 240 ft.

Gunnel – Hagfish – Quillfish

DISTINCTIVE FEATURES: 1. Series of pale blotches at base of dorsal fin. 2. Row of dusky to dark rectangular blotches along side. 3. Dark bar below each eye.
DESCRIPTION: Shades of olive-green to yellow-brown, brown and tan above; yellow to orange or red below. Tiny ventral fins; pectorals relatively large. Continuous dorsal, tail and anal fin; tail is defined by slightly longer rays; anal fin about half the length of dorsal.
ABUNDANCE & DISTRIBUTION: Central California north to Alaska. Also south to Pusan, Korea.
HABITAT & BEHAVIOR: Inhabit muddy substrates and especially in areas of eelgrass beds and leafy algae; also in vicinity of jetties and under docks where they live in jars, cans, tires and other debris. May mix with eelgrass or leafy algae, occasionally hide under rocks or in protective recesses. More commonly in open at night. Usually shallower than 40 feet.
REACTION TO DIVERS: Wary; retreat into tangles of algae or protective recesses when approached. Occasionally curious when peering out protective cover where they can be closely approached with slow nonthreatening movements.

DISTINCTIVE FEATURES: 1. Indistinct head without eyes. 2. Large sucker-like mouth ringed by eight barbels.
DESCRIPTION: Shades of tan to brown or gray; underside pale. No pectoral or ventral fins; dorsal, anal and tail fins continuous. Ten to fourteen gill pores along sides of lower forebody.
ABUNDANCE & DISTRIBUTION: Common (rarely observed) southeast Alaska to southern California. Also south to central Baja.
HABITAT & BEHAVIOR: Burrow into dead, dying or trapped fish, consuming them from the inside. Remain hidden during day; occasionally rest coiled in open at night on muddy, silty, sandy and debris-strewn bottoms.
REACTION TO DIVERS: Remain still; move only if molested.
SIMILAR SPECIES: Lampreys of the family Petromyzontidae have prominent eyes.

DISTINCTIVE FEATURES: Very long, thin body. **1. Small head with large eyes, tiny mouth and protruding lower jaw.**
DESCRIPTION: Somewhat translucent with shades of greenish gray to yellowish or orange. Foredorsal fin composed of tiny spines; long dorsal and anal fins join thread-like tail (often missing).
ABUNDANCE & DISTRIBUTION: Rare Alaska to Oregon.
HABITAT & BEHAVIOR: Swim in snake-like fashion near bottom over a wide range of habitats, including mud and silt to rocky outcroppings and reefs. Often on surface at night where they are attracted to light.
REACTION TO DIVERS: Not particularly shy; allow close view with a slow nonthreatening approach. At night appear mesmerized by light.

Eels & Eel-like Bottom-Dwellers

SADDLEBACK GUNNEL
Pholis ornata
FAMILY:
Gunnel – Pholidae

SIZE: 4 - 9 in., max. 12 in.
DEPTH: 0 - 120 ft.

PACIFIC HAGFISH
Eptatretus stouti
FAMILY:
Hagfish – Myxinidae

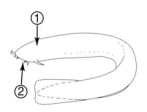

SIZE: 12-20 in., max. 25 in.
DEPTH: 60 - 3,000 ft.

QUILLFISH
Ptilichthys goodei
FAMILY:
Quillfish – Ptilichthyidae

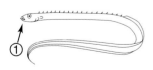

SIZE: 4 - 10 in., max. 13 $\frac{1}{2}$ in.
DEPTH: 0 - 60 ft.

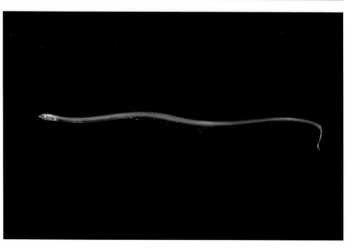

IDENTIFICATION GROUP 4
Elongated Bottom-Dwellers
Fringehead – Kelpfish – Greenling – Others

This ID Group consists of bottom-dwelling fish with elongated bodies and dorsal and anal fins that are obviously separated from the tail.

FAMILY: Clinid — Clinidae
10 Species Included

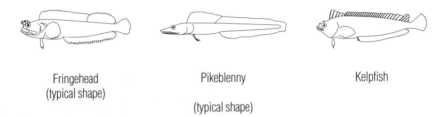

Fringehead
(typical shape)

Pikeblenny

(typical shape)

Kelpfish

Most members of this large family of small fish, commonly known as blennies, inhabit warmer waters. In the chilly waters along the North American Pacific Coast, kelpfish and fringeheads are the most common family members. Ichthyologists also include the region's single pikeblenny species in this family. All have a long single dorsal fin, usually with a tall spinous segment tapering to a long, level, soft rear dorsal that ends with a hump near the tail. Nearly all have scales (This characteristic separates Clinidae from blennies in the Blenniidae family.) and fleshy appendages, called cirri, on their heads.

Blennies generally rest motionlessly in the cover of kelp or on the bottom. A few reside in small holes in rocks. Experts at camouflage, most can change color and/or markings and lighten or darken to blend with their surroundings. Such changes often make identification difficult without careful attention to details.

FAMILY: Greenling — Hexagrammidae
6 Species Included

Greenling
(typical shape)

Longspine Combfish

Lingcod

Greenlings belong to a small family of fish that inhabit the cold waters of the North Pacific. Their single dorsal fin is distinctly notched between the spines in front and soft rays to the rear. They have a long anal fin and usually a square-cut tail. Most have bright, distinctive colors and bold markings, making identification easy. These typically inshore, shallow-water dwellers spend most of their time resting on the bottom propped up by their fins. The Lingcod and Longspine Combfish also belong to the family.

FAMILY: Goby — Gobiidae
4 Species Included

Goby (typical shape)

Gobies make up one of the largest families of marine fish. The majority inhabit tropical reefs; however, a few family members reside in the chilly waters of the North American Pacific Coast. These small fish rarely exceed more than a few inches in length. They have two, distinctly separate, dorsal fins; the foredorsal is supported by spines, while the rear consists of soft rays.

These bottom-dwellers rest on their pectoral and ventral fins. A small suction disc formed between their ventral fins that anchors these small fish in surge or current is a unique family characteristic. Whether resting or swimming, gobies tend to hold their bodies straight and stiff.

FAMILY: Others
13 Species Included

Cod -
Gadidae

Lizardfish -
Synodntidae

Tilefish -
Malacanthidae

Viviparous Brotula -
Bythitidae

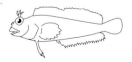

Combtooth Blenny -
Blenniidae

Ronquil -
Bathymasteridae

Fringehead

DISTINCTIVE FEATURES: 1. Two metallic blue spots ringed with yellow on foredorsal fin. 2. Large mouth with jaws extending almost to gill openings.
DESCRIPTION: Shades of brown to gray; may be lightly blotched or barred. Males [pictured] have relatively small cirri over eyes and rear of jaw is yellow to yellowish; females have large cirri over eyes.
ABUNDANCE & DISTRIBUTION: Occasional central and southern California (can be locally abundant). Also south to central Baja.
HABITAT & BEHAVIOR: Inhabit hard sand and mud bottoms along exposed coastlines beyond breakers. Occupy crevices, holes, burrows, empty shells, clams, bottles and cans. Aggressive toward all intruders. Males vigorously guard egg masses.
REACTION TO DIVERS: Fearless; will charge, snap at and occasionally bite all close approaching intruders, including divers.

DISTINCTIVE FEATURES: 1. One blue spot ringed with yellow on foredorsal fin. 2. Three pairs of branched cirri around upper-forward edge of eye; the first is largest and longest, and can be quite dramatic. 3. Large mouth, jaw extends beyond eye.
DESCRIPTION: Shades of brown with numerous white, black and occasionally blue or red flecks; may be mottled, blotched or barred.
ABUNDANCE & DISTRIBUTION: Occasional California. Also south to northern Baja.
HABITAT & BEHAVIOR: Inhabit wide range of bottom environments from coastlines to jetties, inlets, bays and in vicinity of docks and piers. Occupy crevices, holes, burrows, empty shells, clams, bottles and cans. Often rest in opening with only head exposed; rarely leave protective cover of habitation. Aggressive toward all intruders. Both males and females vigorously guard orangish egg masses.
REACTION TO DIVERS: Fearless; will charge, snap at and occasionally bite all close approaching intruders, including divers.

Onespot Fringehead
Color variation.

Elongated Bottom-Dwellers

SARCASTIC FRINGEHEAD
Neoclinus blanchardi
FAMILY:
Clinid – Clinidae

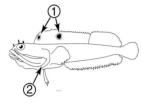

SIZE: 3-8 in., max. 1 ft.
DEPTH: 10 - 240 ft.

ONESPOT FRINGEHEAD
Neoclinus uninotatus
FAMILY:
Clinid – Clinidae

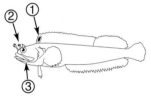

SIZE: 3 - 7 ½ in., max. 10 in.
DEPTH: 10 - 100 ft.

Onespot Fringehead
First pair of cirri extremely elongated.

Fringehead–Pikeblenny

DISTINCTIVE FEATURES: 1. Three pair of massive, multi-branched cirri around upper-front edge of eye.
DESCRIPTION: Uniform or mottled, blotched and barred in shades of red to orange, brown, olive and gray; often with numerous white to blue specks. Change color and markings to match surroundings. Dorsal fin uniform in height and without ocellated spots.
ABUNDANCE & DISTRIBUTION: Common to occasional Monterey to southern California. Also south to central Baja.
HABITAT & BEHAVIOR: Inhabit rocky coastlines, reefs and outcroppings; also in bays and along jetties. Occupy small holes, empty shells and bottles. Often rest in opening with only head exposed; rarely leave protective cover.
REACTION TO DIVERS: Remain in opening; do not retreat into hole unless molested. Often appear curious, peering out at diver.

DISTINCTIVE FEATURES: 1. Throat usually orange. 2. Pointed snout. 3. Long mouth extends beyond eye.
DESCRIPTION: Shades of brown to brownish green with whitish and blue specks and spots; darkish blotches on sides; often ocellated spot between first two rays of dorsal fin. Male's [pictured] dorsal fin is dark and somewhat taller in front: female's dorsal fin is relatively even in height and mostly translucent. Scaleless body.
ABUNDANCE & DISTRIBUTION: Occasional southern California. Also Baja and Gulf of California.
HABITAT & BEHAVIOR: Inhabit shallow sandy and mud bottoms; live in parchment-like tubes of marine worms. May rest with only head extending from tube, but often move about in surrounding area. Territorial, aggressively guard tube.
REACTION TO DIVERS: Not shy; usually can be closely viewed with slow nonthreatening approach.

Elongated Bottom-Dwellers

YELLOWFIN FRINGEHEAD
Neoclinus stephensae
FAMILY:
Clinid – Clinidae

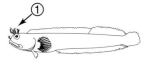

SIZE: 1½ - 3 in.,
max. 4 in.
DEPTH: 10 - 100 ft.

Yellowfin Fringehead
Typical habitation.
Color variations.
[left and right]

ORANGETHROAT PIKEBLENNY
Chaenopsis alepidota
FAMILY:
Clinid – Clinidae

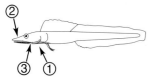

SIZE: 2 - 4 in.,
max. 6 in.
DEPTH: 10 - 40 ft.

Kelpfish

DISTINCTIVE FEATURES: 1. Elongated head with pointed, slightly upturned, snout. 2. Forked tail. (All other kelpfish have rounded tails.)
DESCRIPTION: Color and markings vary greatly, including shades of yellow, yellow-green, green, green-brown, red-brown, brown and lavender; can be unmarked, blotched, spotted or barred. Rapidly change color and markings to camouflage with background. Spinous foredorsal fin taller than soft dorsal.
ABUNDANCE & DISTRIBUTION: Common southern California; occasional, becoming rare, north to British Columbia, Canada.
HABITAT & BEHAVIOR: Inhabit kelp beds and areas of luxuriant leafy algae growth. Nestle in blades, blending almost perfectly with background.
REACTION TO DIVERS: Shy; quickly move away when obviously observed; however, apparently relying on camouflage, usually move only when closely approached.

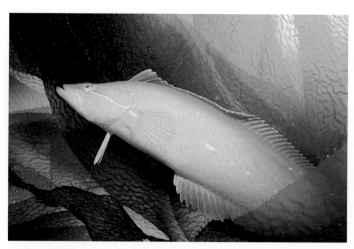

Giant Kelpfish
Yellow variation.

DISTINCTIVE FEATURES: 1. A single row of ocellated spots (occasionally obscure) along upper side; the first and those toward the rear tend to be more obvious. 2. Soft rays of raised rear dorsal fin become more widely spaced toward rear. 3. Short pectoral fin does not reach beginning of anal fin.
DESCRIPTION: In shallow water, shades of brown to reddish brown, gray or lavender; below 65 feet, usually shades of red, scarlet and pink. May be relatively uniform in color or striped or blotched and spotted. Tail fin rounded.
ABUNDANCE & DISTRIBUTION: Occasional BC to southern California. Also south to northern Baja.
HABITAT & BEHAVIOR: Inhabit rough rocky areas with numerous recesses and abundant algae growth. Commonly intertidal or quite shallow in northern extent of range, tend to live in deeper water to the south, rarely above 60 feet in southern California. Rest on bottom, blending with surroundings.
REACTION TO DIVERS: Not shy; remain still, apparently relying of camouflage.
SIMILAR SPECIES: Striped Kelpfish, *G. metzi*, distinguished by evenly spaced rays of raised rear dorsal fin, absence of ocellated spots and often striped.
NOTE: Red, deep-dwelling specimens [pictured] formerly classified as a separate species, Scarlet Kelpfish, *G. erythra*.

Elongated Bottom-Dwellers

GIANT KELPFISH
Heterostichus rostratus
FAMILY:
Clinid – Clinidae

SIZE: 6-16 in.,
max. 2 ft.
DEPTH: 3-132 ft.

Giant Kelpfish Juvenile

CREVICE KELPFISH
Gibbonsia montereyensis
FAMILY:
Clinid – Clinidae

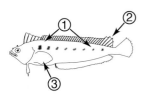

SIZE: 2 1/2 - 4 1/2 in.,
max. 6 in.
DEPTH: 0 - 120 ft.

101

Blenny-Kelpfish

DISTINCTIVE FEATURES: 1. Thin red stripes on back gradating to orange near midbody and yellow below. 2. Rays of raised rear dorsal fin evenly spaced. 3. Long pectoral fin extends beyond beginning of anal fin.

DESCRIPTION: Whitish to gray, tan or olive undercolor; red to orange and occasionally blue markings on head; tail barred with red and white; several darkish blotches on sides. Foredorsal fin is not taller than soft dorsal.

ABUNDANCE & DISTRIBUTION: Occasional southern California. Also south to central Baja.

HABITAT & BEHAVIOR: Inhabit deep rocky reefs and outcroppings. Perch on bottom blending with substrate.

REACTION TO DIVERS: Wary; but usually allow a slow, close nonthreatening approach before bolting to new perch.

NOTE: Also comonly known as "Deepwater Kelpfish."

DISTINCTIVE FEATURES: 1. Foredorsal fin same height as soft dorsal and first few spines have flexible bent tips. 2. Rays of raised rear dorsal fin evenly spaced. 3. Pale rounded blotch extends onto cheek from lower rear quarter of eye.

DESCRIPTION: Blotched, barred and striped in shades of red to maroon, lavender, orange, tan, brown and occasionally greenish; tiny white to pale blue spots cover body; usually a row of dark blotches on upper body. Occasionally greenish spot on foredorsal fin. Long pectoral fin extends beyond beginning of anal fin. Change color and markings to match background.

ABUNDANCE & DISTRIBUTION: Abundant around islands of southern California; occasional coastal areas of southern California. Also south to central Baja.

HABITAT & BEHAVIOR: Inhabit rocky coastlines, outcroppings and reefs with abundant seaweed and algae growth; also in kelp beds. Perch on substrate mixed in with growth, changing color and markings to camouflage with surroundings.

REACTION TO DIVERS: Wary; but usually allow a slow, close nonthreatening approach before bolting to resettle nearby.

Island Kelpfish
Reddish variation.

Elongated Bottom-Dwellers

DEEPWATER BLENNY
Cryptotrema corallinum
FAMILY:
Clinid – Clinidae

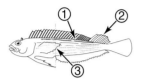

SIZE: 2 - 4 in.,
max. 5 in.
DEPTH: 70 - 300 ft.

ISLAND KELPFISH
Alloclinus holderi
FAMILY:
Clinid – Clinidae

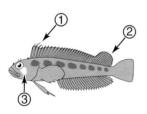

SIZE: 2 - 3 ½ in.,
max. 4 in.
DEPTH: 0 - 160 ft.

Island Kelpfish
Lavender variation.

103

Kelpfish–Greenling

DISTINCTIVE FEATURES: 1. One to three dark ocellated spots on back; commonly two, one above pectoral fin and the other on rear body. 2. Rays of raised rear dorsal fin evenly spaced. 3. Pectoral fin extends almost to beginning of anal fin.

DESCRIPTION: Barred, blotched and striped shades of gray to tan, brown, maroon and green; change color and markings to match background. Spinous foredorsal fin raised taller than soft dorsal.

ABUNDANCE & DISTRIBUTION: Occasional central and southern California. Also south to southern Baja.

HABITAT & BEHAVIOR: Inhabit rocky coastlines, outcroppings and reefs with abundant algae and seaweed growth. Perch on bottom mixed in with growth, changing color and markings to camouflage with surroundings. Males guard white egg masses attached to growth.

REACTION TO DIVERS: Not shy; remain still, apparently relying on camouflage.

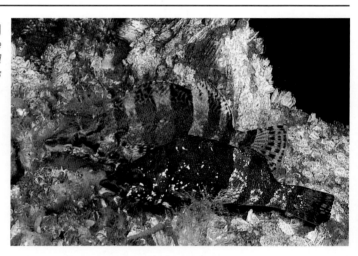

Painted Greenling
Mating couple; male is nearly black and female's bars are brown.

DISTINCTIVE FEATURES: 1. Five to six, bold, dark (usually red) bars encircle fins and body. 2. Pointed snout. 3. Two pair of cirri — one above each eye and another midway between eyes and dorsal fin.

DESCRIPTION: Undercolor white to pinkish, cream, and light brown; red bars often shaded with brown. Two bands radiate back from eye and another forward. Single lateral line. Mature males often turn nearly black during winter mating season and the bars of females may become brown.

ABUNDANCE & DISTRIBUTION: Occasional southern California north to British Columbia; rare to Bering Sea, Alaska. Also south to central Baja.

HABITAT & BEHAVIOR: Inhabit shallow rocky areas, especially in tidal channels; also in vicinity of docks. Hover in water column just above bottom or may actively move from perch to perch. Occasionally rest on *Telia* anemones, apparently immune to their sting. Mate in winter; males aggressively guard egg masses.

REACTION TO DIVERS: Unafraid and occasionally curious; may follow diver. When guarding eggs males appear fearless and often charge and nip at all intruders, including divers.

NOTE: Also commonly known as "Convict Fish."

Elongated Bottom-Dwellers

SPOTTED KELPFISH
Gibbonsia elegans
FAMILY:
Clinid – Clinidae

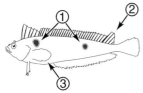

SIZE: 2½ - 4½ in.,
max. 6½ in.
DEPTH: 0 - 190 ft.

Spotted Kelpfish
Maroon variation.

PAINTED GREENLING
Oxylebius pictus
FAMILY:
Greenling –
Hexagrammidae

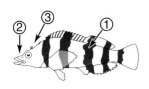

SIZE: 4 - 6 in.,
max. 10 in.
DEPTH: 3 - 160 ft.

Greenling

DISTINCTIVE FEATURES: 1. A large, bushy cirri above each eye. 2. Prominent diagonal pale band from eye toward pectoral; additional bands may also radiate below and behind eyes.
DESCRIPTION: Vary greatly; usually dark olive to greenish brown to brown undercolor with blotches of white to tan or turquoise; bright red in larger males. Four lateral lines.
ABUNDANCE & DISTRIBUTION: Occasional Bering Sea, Alaska, to central California; uncommon to rare southern California.
HABITAT & BEHAVIOR: Inhabit rocky areas of surge-swept shorelines with dense growths of kelp. Lurk in shadows blending with background, rarely leave cover of kelp.
REACTION TO DIVERS: Extremely wary; usually bolt when approached. Occasionally a very slow nonthreatening approach will allow a closer view.

DISTINCTIVE FEATURES: MALE: 1. Blue irregular spots on head and forebody, outlined by a few small, dark reddish brown spots. **FEMALE:** 2. Speckled with red-brown to gold spots over bluish white to pale cream, light brown or gray undercolor. (Similar Whitespotted Greenling [next] distinguished by numerous white spots.)
DESCRIPTION: MALE: Olive to brown or gray to bluish gray. Pair of small cirri above eyes and another tiny pair between eyes and dorsal fin. Five lateral lines.
ABUNDANCE & DISTRIBUTION: Abundant to common Aleutian Islands, Alaska, to Washington; occasional Oregon to central California; rare southern California.
HABITAT & BEHAVIOR: Generally inhabit kelp beds, but also around rocky areas and on sand bottoms.
REACTION TO DIVERS: Unafraid and curious; may follow diver. Can often be hand-fed.

Kelp Greenling Female
Color and marking variation.

Elongated Bottom-Dwellers

ROCK GREENLING
Hexagrammos lagocephalus
FAMILY:
Greenling –
Hexagrammidae

SIZE: 10 - 18 in.,
max. 2 ft.
DEPTH: 0 - 60 ft.

KELP GREENLING
Hexagrammos decagrammus
Male
FAMILY:
Greenling –
Hexagrammidae

SIZE: 10 - 18 in., max. 2 ft.
DEPTH: 0 - 150 ft.

Kelp Greenling Female
Typical color and markings.

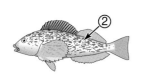

Greenling

DISTINCTIVE FEATURES: 1. Numerous white spots over body. (Similar female Kelp Greenling [previous] distinguished by numerous dark spots and occasionally a few larger white spots.)
DESCRIPTION: Shades of light brown to greenish brown, brown or red-brown. Small pair of cirri, one over each eye. Four lateral lines — three long, one short.
ABUNDANCE & DISTRIBUTION: Occasional Aleutian Islands, Alaska, to Puget Sound, Washington; uncommon south to southern Oregon.
HABITAT & BEHAVIOR: Wide range of inshore habitats, including sand plains, eelgrass beds, rocky areas with marine plants and algae, and in vicinity of docks. Active swimmers. Mate in winter; males aggressively guard egg masses.
REACTION TO DIVERS: Wary; however, a slow nonthreatening approach usually allows a close view. When guarding eggs males appear fearless and often charge and nip at intruders, including divers.

DISTINCTIVE FEATURES: 1. Large mouth with prominent canine teeth. 2. Long, even spinous dorsal separated by notch just before taller soft rear dorsal.
DESCRIPTION: Numerous dark spots and several darkish blotches over cream to tan, light brown, gray or black undercolor, belly white; may have greenish or bluish tint on back. Lighten, darken and change color to blend with background. Single, prominent, whitish lateral line.
ABUNDANCE & DISTRIBUTION: Occasional Bering Sea, Alaska, to northern Baja.
HABITAT & BEHAVIOR: Inhabit rocky areas. Rest on bottom or patrol established territory. After mating, males guard masses of white eggs. Older females larger than males.
REACTION TO DIVERS: Wary; generally remain still but move away when closely approached. When guarding eggs males appear fearless and often charge and nip at all intruders, including divers.

Lingcod
Color variation, note red spots.

Elongated Bottom-Dwellers

WHITESPOTTED GREENLING
Hexagrammos stelleri
FAMILY:
Greenling –
Hexagrammidae

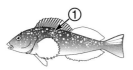

SIZE: 6 -12 in.,
max. 19 in.
DEPTH: 3 -150 ft.

LINGCOD
Ophiodon elongatus
FAMILY:
Greenling –
Hexagrammidae

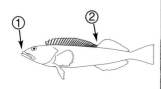

SIZE: 1½ - 3½ ft.,
max. 5 ft.
DEPTH: 6 -1,400 ft.

Lingcod
Black variation.

109

Greenling–Cod

DISTINCTIVE FEATURES: 1. Exceptionally long second dorsal fin spine. 2. Dark stripe extends from snout through eye. (Similar Spotfin Sculpin and Threadfin Sculpin [pg. 69] lack this stripe.)
DESCRIPTION: Shades of yellow-brown to green-brown. About five dusky saddles across back and a scattering of spots over body; red and gold ocellated spots on dorsal fin.
ABUNDANCE & DISTRIBUTION: Rare Vancouver Island, Canada, to southern California. Also south to central Baja.
HABITAT & BEHAVIOR: Inhabit areas of sand, silt and mud. Generally rest motionless on bottom.
REACTION TO DIVERS: Wary; usually remain still, but bolt when closely approached. At night easily approached, apparently mesmerized by diver's handlight.
NOTE: Formerly classified in the family Zaniolepididae, commonly known as combfishes.

DISTINCTIVE FEATURES: 1. Three separate dorsal fins. 2. Long chin barbel, "whisker," longer than diameter of eye. (Similar Walleye Pollock [next] has a slightly projecting lower jaw and minute or no chin barbel.)
DESCRIPTION: Elongated, silvery gray to brown body with brown spots or scrawl markings. Two anal fins, the first below second dorsal. Tail square-cut.
ABUNDANCE & DISTRIBUTION: Common Washington, British Columbia, Canada and southeast Alaska; occasional Oregon to northern California and Gulf of Alaska; uncommon to rare to southern California. Also to Japan, Korea and China.
HABITAT & BEHAVIOR: Schools and occasionally individuals cruise just above soft bottoms and areas of gravel and rocky rubble while searching for prey. Rarely shallower than 40 feet, tend to be deeper in fall and winter.
REACTION TO DIVERS: Shy; schools generally move away when approached. Occasionally individuals may be approached with slow nonthreatening movements.
SIMILAR SPECIES: Pacific Tomcod, *Microgadus proximus*, distinguished by short chin barbel (less than diameter of eye).

DISTINCTIVE FEATURES: 1. Three separate dorsal fins. 2. Slightly projecting lower jaw and minute or no chin barbel. (Similar Pacific Cod [previous] distinguished by long chin barbel.)
DESCRIPTION: Elongated, silvery body with brownish to greenish lightly mottled back, gradating to silver on sides and whitish belly. Two anal fins, the first below second dorsal. Tail square-cut. Young have two or three narrow yellowish stripes on sides.
ABUNDANCE & DISTRIBUTION: Abundant Gulf of Alaska; common British Columbia; occasional south to central California. Also to Japan and Korea.
HABITAT & BEHAVIOR: Schools of juveniles or occasionally individuals cruise just above shallow soft bottoms and areas of gravel and rocky rubble in search of prey. Adults most common below safe diving limits, school both in mid-water and near bottom; at night individuals occasionally forage near bottom in shallow water.
REACTION TO DIVERS: Shy; schools generally move away when approached. Occasionally individuals may be approached with slow nonthreatening movements.
SIMILAR SPECIES: Pacific Hake, *Merluccius productus*, distinguished by no chin barbel and only two dorsal fins; the second is deeply notched, but not separated.

Elongated Bottom-Dwellers

LONGSPINE COMBFISH
Zaniolepis latipinnis
FAMILY:
Greenling –
Hexagrammidae

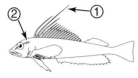

SIZE: 5 - 8 in.,
max. 1 ft.
DEPTH: 60 - 600 ft.

PACIFIC COD
Gadus macrocephalus
FAMILY:
Cod – Gadidae

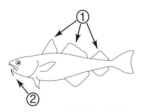

SIZE: 1 - 3 ft.,
max. 3 3/4 ft.
DEPTH: 3 - 3,000 ft.

WALLEYE POLLOCK
Theragra chalcogramma
FAMILY:
Cod – Gadidae

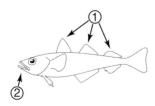

SIZE: 1 - 2 1/2 ft.,
max. 3 ft.
DEPTH: 3 - 3,200 ft.

Lizardfish–Tilefish–Brotula

DISTINCTIVE FEATURES: Tapering cylindrical body. **1. Pointed snout with long slanting mouth that extends beyond eye.**
DESCRIPTION: Upper body mottled and in patterned shades of brown becoming pale on lower sides and belly. Small, fleshy second dorsal fin without spines or rays; deeply forked tail. Numerous, short needle-like teeth.
ABUNDANCE & DISTRIBUTION: Occasional (but uncommonly observed because of habit of burying in bottom material) southern and central California. Also south to Baja and northern Gulf of California.
HABITAT & BEHAVIOR: Inhabit mud and sand flats, most common between 50-130 feet. Rest on or bury in bottom material, often with only eyes exposed.
REACTION TO DIVERS: Remain motionless, apparently relying on camouflage. Bolt only when very closely approached or molested.

DISTINCTIVE FEATURES: 1. Yellowish borders on tail. 2. Yellow streak down center of pectoral fin.
DESCRIPTION: Elongated, heavy body with small terminal mouth. Back silvery light gray to brown; belly white; may have bluish tints. Fins often yellowish and may have bluish stripes.
ABUNDANCE & DISTRIBUTION: Common southern California; occasional central California; uncommon to rare northern California to Vancouver Island, Canada. Also south to Peru.
HABITAT & BEHAVIOR: Inhabit rocky areas, especially high profile reefs, and occasionally kelp beds. Most common between 80 and 180 feet around offshore islands. Often gather in aggregations, drifting above bottom. May dig in soft bottom material for food.
REACTION TO DIVERS: Tend to ignore divers, but move away when closely approached. Slow nonthreatening movements may allow close view.

DISTINCTIVE FEATURES: Shades of brownish red. **1. Four, (two long, two short) barbel-like, ventral fin rays. 2. Dorsal and anal fins extend to base of tail, where deep notches set off the small tail.** (Similar Purple Brotula [next] distinguished by two ventral fin rays, and continuous dorsal, anal and tail fins end with a rounded point.)
DESCRIPTION: Bright red to red and occasionally orangish red to brownish red; lower side and belly somewhat paler. Fore-lateral line arched with slight break between it and straight rear-lateral line that extends to tail.
ABUNDANCE & DISTRIBUTION: Occasional (but rarely observed because of secretive nature) southeastern Alaska to southern California. Also northern Baja.
HABITAT & BEHAVIOR: Inhabit rocky, boulder-strewn areas, reefs and walls. Haunt crevices, caves and other recesses, occasionally hovering in entrance. Rarely shallower than 60 feet.
REACTION TO DIVERS: Shy; retreat deep into recess when approached. Slow nonthreatening movements may allow close view.

Elongated Bottom-Dwellers

CALIFORNIA LIZARDFISH
Synodus lucioceps
FAMILY:
Lizardfish – Synodontidae

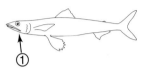

SIZE: 6 - 15 in.,
max. 25 in.
DEPTH: 6 - 750 ft.

OCEAN WHITEFISH
Caulolatilus princeps
FAMILY:
Tilefish – Malacanthidae

SIZE: 8 - 15 in.,
max. 1 1/2 ft.
DEPTH: 4 - 450 ft.

RED BROTULA
Brosmophycis marginata
FAMILY:
Viviparous Brotula –
Bythitidae

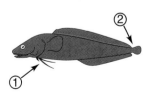

SIZE: 10 - 15 in.,
max. 1 1/2 ft.
DEPTH: 10 - 800 ft.

Brotula–Combtooth Blenny

DISTINCTIVE FEATURES: Dark purple to purplish gray or black. **1. Two, short, barbel-like ventral fin rays. 2. Continuous dorsal, anal and tail fins end with a rounded point.** (Similar Red Brotula [previous] distinguished by four ventral fin rays and tail separated from fins.)

DESCRIPTION: Occasionally reddish purple. Two lateral lines; the upper extends from head and runs about three-quarters the length of upper body, the lower begins at midbody above the beginning of anal fin and runs to tail.

ABUNDANCE & DISTRIBUTION: Occasional (but rarely observed because of secretive nature) southern California. Also south to Panama and Galapagos.

HABITAT & BEHAVIOR: Inhabit rocky, boulder-strewn areas, reefs and walls.
Haunt crevices, caves and other recesses, occasionally hovering near entrance during day; may venture into open at night.

REACTION TO DIVERS: Shy; nearly always retreat deep into recess when approached. Stalking with slow nonthreatening movements may allow close view.

DISTINCTIVE FEATURES: 1. Cirri above eyes branch at or near base. 2. Eight dark saddles across back.

DESCRIPTION: Shades of brown to brownish green or gray. Often spotted or marked with red, especially on head; frequently two bands below eyes. Indention behind eyes give profile a notched appearance.

ABUNDANCE & DISTRIBUTION: Occasional southern California. Also south to southern Baja.

HABITAT & BEHAVIOR: Inhabit tide pools, rocky areas and in vicinity of docks and pilings. Often live in barnacle shells and other protective recesses [see above] where they perch in opening with only their heads exposed. Males guard egg clusters.

REACTION TO DIVERS: Not shy; usually can be closely viewed with slow nonthreatening approach.

NOTE: Also commonly known as "Notchbrow Blenny."

SIMILAR SPECIES: Bay Blenny, *H. gentilis*, distinguished by unbranched cirrus with serrated rear edge above each eye. Central and southern California. Mussel Blenny, *H. jenkinsi*, distinguished by cirrus branched only near tip above each eye. Neither species has an indention behind eyes.

Elongated Bottom-Dwellers

PURPLE BROTULA
Oligopus diagrammus
FAMILY:
Vivaporus Brotula –
Bythitidae

SIZE: 4-6 in.,
max. 8 in.
DEPTH: 15-75 ft.

Purple Brotula
Black variation.
[right]

Rockpool Blenny [left]
Typically live in holes or empty barnacle shells where they perch with only their heads exposed.

ROCKPOOL BLENNY
Hypsoblennius gilberti
FAMILY:
Combtooth Blenny –
Blenniidae

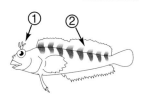

SIZE: 2-4 in.,
max. 6 3/4 in.
DEPTH: 0-60 ft.

115

Ronquil–Goby

DISTINCTIVE FEATURES: 1. Two rows of prominent white squarish to rectangular spots on sides below high lateral line.
DESCRIPTION: Tan to brown undercolor; row of dark blotches along base of long dorsal fin and another along midbody; anal fin blue and yellow-striped. Pectoral fin does not extend to start of anal fin.
ABUNDANCE & DISTRIBUTION: Uncommon San Francisco to southern California. Also south to northern Baja.
HABITAT & BEHAVIOR: Inhabit rocky areas along exposed coastlines, deep rocky outcroppings and kelp forests. Rest on bottom blending with substrate.
REACTION TO DIVERS: Not shy; a slow nonthreatening approach usually allows a close view

---

DISTINCTIVE FEATURES: 1. Pale orangish spots or stripes below eye. 2. Darkish patch between eyes and nape. 3. Single, long, continuous straight dorsal fin. (Similar Blackeye Goby [next] and Bay Goby [top, following page] are distinguished by two dorsal fins.)
DESCRIPTION: Vary from pale to darker shades of orangish cream to brown, olive-green and gray; may have vague bars on back and sides. High, straight lateral line. Pectoral fin extends to start of anal fin. Fore-rays of dorsal fin unbranched, rear rays forked. In late winter/early spring mature males display mating colors of purple blotches on head and bright blue and yellow anal fins.
ABUNDANCE & DISTRIBUTION: Occasional Bering Sea, Alaska, to Monterey Bay, California.
HABITAT & BEHAVIOR: Inhabit sandy, silty areas strewn with rocks and rocky outcroppings; also in vicinity of docks and jetties, often live in man-made litter such as bottles and cans. Rest on bottom perched on pectoral fins near opening to protective recesses. Generally in deeper water from California to Washington and shallower to the north.
REACTION TO DIVERS: Wary; retreat into recess when approached; apparently curious, often return to opening to peer out. Slow nonthreatening approach may allow close view.

DISTINCTIVE FEATURES: 1. Black eye. 2. Black edge on foredorsal fin. (Similar Bay Goby [next] distinguished by lack of black eye. Similar Northern Ronquil [previous] distinguished by single dorsal fin.)
DESCRIPTION: Dark to pale tan.
ABUNDANCE & DISTRIBUTION: Common southern California to central British Columbia; uncommon north to near Alaska. Also south to central Baja.
HABITAT & BEHAVIOR: Inhabit sandy areas near rocky outcroppings, reefs and in vicinity of docks. Live in protective recesses, under rocks, or excavate dens in sand or silt; may occupy man-made litter, such as cans, jars and tires. Rest on bottom near home, blending with background. During mating season often make peculiar open-mouth displays.
REACTION TO DIVERS: Remain still, apparently relying on camouflage. Dart to protective recess when closely approached.

Elongated Bottom-Dwellers

STRIPEDFIN RONQUIL
Rathbunella hypoplecta
FAMILY:
Ronquil – Bathymasteridae

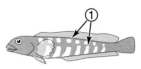

SIZE: 3 - 5 in.,
max. 6 1/2 in.
DEPTH: 15 - 300 ft.

NORTHERN RONQUIL
Ronquilus jordani
FAMILY:
Ronquil – Bathymasteridae

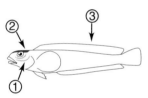

SIZE: 4 - 6 in.,
max. 7 in.
DEPTH: 10 - 540 ft.

BLACKEYE GOBY
Coryphopterus nicholsi
FAMILY:
Goby – Gobiidae

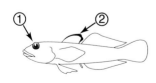

SIZE: 1 1/2 - 4 in.,
max. 6 in.
DEPTH: 0 - 340 ft.

Goby

DISTINCTIVE FEATURES: 1. Black edge on foredorsal fin. (Similar Blackeye Goby [previous] distinguished by black eye.)
DESCRIPTION: Tan to reddish brown, brown, greenish brown or olive; frequently somewhat translucent. Usually dark streak extends from lower front of eye to jaw; often with darkish body blotches or large dash markings.
ABUNDANCE & DISTRIBUTION: Common southern California to northern British Columbia. Also south to central Baja.
HABITAT & BEHAVIOR: Inhabit flat sand, mud and silty bottoms. Rest on bottom near small hole, blending with background.
REACTION TO DIVERS: Wary, but remain still, apparently relying on camouflage. Dart into hole when closely approached.

DISTINCTIVE FEATURES: Brilliant red. **1. Four to nine electric-blue bars.** (Similar Zebra Goby [next] distinguished by more numerous and thinner bars.)
DESCRIPTION: Foredorsal fin tall, especially in males. Rear body bars thinner.
ABUNDANCE & DISTRIBUTION: Abundant to common southern and south central California. Also south to central Baja and Gulf of California.
HABITAT & BEHAVIOR: Inhabit open rocky areas. Territorial; perch in open, retreating to nearby protective recess only when threatened. Males guard eggs laid by females.
REACTION TO DIVERS: Wary, but generally remain still, allowing slow nonthreatening approach. Dart for cover only when very closely approached.

DISTINCTIVE FEATURES: Brilliant red. **1. Numerous thin, bright blue bands from head to tail with thinner blue to dusky bars between.** (Similar Bluebanded Goby [previous] distinguished by only a few wide bars.)
DESCRIPTION: Occasionally reddish orange to brilliant orange. Foredorsal fin tall, especially males.
ABUNDANCE & DISTRIBUTION: Occasional central and southern California. Also Baja and Gulf of California.
HABITAT & BEHAVIOR: Cryptic, inhabit rocky areas, often in cracks, crevices, caves and other recesses; occasionally take cover in protective spines of sea urchins. Males are territorial during mating season and aggressively guard eggs.
REACTION TO DIVERS: Shy; usually dart deeper into recess when approached. Males can easily be approached when guarding eggs.

Elongated Bottom-Dwellers

BAY GOBY
Lepidogobius lepidus
FAMILY:
Goby – Gobiidae

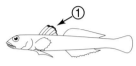

SIZE: 1½ - 3 in.,
max. 4 in.
DEPTH: 0 - 660 ft.

BLUEBANDED GOBY
Lythrypnus dalli
FAMILY:
Goby – Gobiidae

SIZE: ¾ - 2 in.,
max. 2½ in.
DEPTH: 0 - 250 ft.

ZEBRA GOBY
Lythrypnus zebra
FAMILY:
Goby – Gobiidae

SIZE: ¾ - 2 in.,
max. 2¼ in.
DEPTH: 0 - 300 ft.

IDENTIFICATION GROUP 5
Flatfish/Bottom-Dwellers
Flounder – Turbot – Sole – Halibut – Sanddab

This ID Group consists of flattened fish that rest on the bottom on either their right or left side.

FAMILY: Righteye Flounder — Pleuronectidae
Lefteye Flounder — Bothidae
15 Species Included

Righteye Flounder (typical shape)

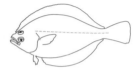

Lefteye Flounder (typical shape)

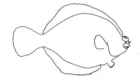

Hornyhead Turbot

Slender Sole

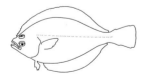

Pacific Sanddab

California Halibut

Flounders are unique, flat fish that actually lie on their sides. Those in the lefteye flounder family usually rest on their flattened right side with both eyes on the exposed left side. Righteye flounders are the reverse. Nearly all species from both families reside in cool to cold waters, but, even though common along the California to Alaska coastline, only a few are regularly observed by divers.

A few weeks after birth, and just before settling to the bottom, the bilaterally symmetrical pelagic larval stage undergoes an amazing transformation. The skull twists and one eye migrates through a slit in the head and settles adjacent to the other eye on what becomes the exposed side of the now flat fish. The visible side develops a pigment pattern while the underside is whitish. The exposed pectoral fin, which resembles a dorsal fin, remains centrally located while the dorsal and anal fins line the edges of the flattened and somewhat circular fish. The eyes protrude noticeably, sometimes appearing to be raised on short, thick stalks. When swimming, they glide over the bottom with a undulating motion.

Flounders are masters of camouflage, changing their color and markings to match the substrate. They often enhance this deception by partially burying themselves in soft bottom material with only their eyes exposed. Flounders are difficult to distinguish, but, with careful attention to detail, most can be identified to species.

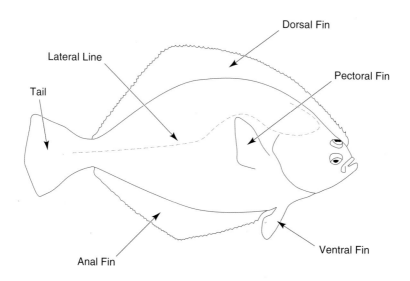

Righteye Flounder

DISTINCTIVE FEATURES: 1. Short ridge between eyes with a prominent spine at each end (front spine occasionally missing). 2. Large protruding eyes near tip of snout; mouth small.
DESCRIPTION: Shades of brown with dark blotches and white spots. Thin, dark outline around scales forms honeycomb pattern. Foredorsal fin (typically the first six or less rays) curves under toward blind side. Lateral line nearly straight with long curving branch from head to near mid-dorsal fin. Oval body; rounded tail.
ABUNDANCE & DISTRIBUTION: Common southern California; occasional to uncommon central and northern California. Also south to southern Baja and northern Gulf of California.
HABITAT & BEHAVIOR: Inhabit flat sand, silt and mud bottoms. Rest on bottom blending with substrate. Actively forage during day. Generally below 35 feet.
REACTION TO DIVERS: Apparently relying on camouflage, remain still. Bolt only when closely approached or molested.

DISTINCTIVE FEATURES: 1. Dark crescent-shaped marking on tail near base followed by large spot form "C-O." 2. Usually dark prominent midbody spot with pale patch behind. 3. Large, protruding bulbous eyes (source of alternate common name, "popeye") near tip of snout.
DESCRIPTION: Mottled, spotted and blotched in rich shades of brown with occasional white spots. May lighten, darken and/or display patches of other colors to blend with substrate. Foredorsal fin (typically first six or less rays) curves under toward blind side. Lateral line nearly straight with long curving branch from head to near mid-dorsal fin. Oval body; broad rounded tail.
ABUNDANCE & DISTRIBUTION: Occasional southern California to southeast Alaska. Also south to northern Baja.
HABITAT & BEHAVIOR: Inhabit flat sandy, silty or mud bottoms; often in vicinity of rocky outcroppings, reefs and eelgrass beds. Rest on bottom, frequently partially or completely covered with soft bottom material; occasionally rest on or glide slowly over eelgrass and algae blades. Generally shallower than 50 feet.
REACTION TO DIVERS: Remain still; bolt only when closely approached.
NOTE: Also commonly known as "Popeye."

C-O Sole
Color variation; note greenish tints displayed to enhance camouflage.

Flatfish/Bottom-Dwellers

HORNYHEAD TURBOT
Pleuronichthys verticalis
FAMILY:
Righteye Flounder –
Pleuronectidae

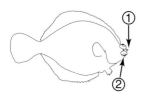

SIZE: 6-11 in.,
max. 15 in.
DEPTH: 20-700

C-O SOLE
Pleuronichthys coenosus
FAMILY:
Righteye Flounder –
Pleuronectidae

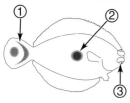

SIZE: 6-12 in.,
max. 14½ in.
DEPTH: 3-1,200 ft.

C-O Sole
Color variation; note patches of pastel lavender and yellow displayed to enhance camouflage.

Righteye Flounder

DISTINCTIVE FEATURES: 1. Foredorsal fin (typically the first 9-12 rays) curls out from underside. 2. Short ridge between eyes with a small conical horn at each end. 3. Large protruding eyes near tip of snout.
DESCRIPTION: Shades of brown, occasionally with some mottling. Broad, rounded unpatterned tail. Lateral line nearly straight with long curving branch from head to near center of dorsal fin. Oval body; middle of dorsal and anal fins wide.
ABUNDANCE & DISTRIBUTION: Occasional California; uncommon Oregon to Alaska. Also south to central Baja.
HABITAT & BEHAVIOR: Inhabit flat sand, silt and mud bottoms Rest on bottom blending with substrate; often bury, leaving only eyes exposed. Generally below 70 feet.
REACTION TO DIVERS: Apparently relying on camouflage, remain still. Bolt only when closely approached or molested.

DISTINCTIVE FEATURES: 1. Rough scales with thin dark borders cover upper side. 2. Lateral line arches steeply over pectoral fin with a short branching lateral line running from head along foredorsal fin.
DESCRIPTION: Mottled and blotched shades of brown, occasionally gray. Oval body; tail nearly square-cut.
ABUNDANCE & DISTRIBUTION: Occasional Southern California to Alaska. Also to Japan and Korea.
HABITAT & BEHAVIOR: Prefer flat gravel bottoms, also on sand, silt and mud. Unlike other flatfish, Rock Soles rest on bottom propped up on their fins; seldom lie flat and rarely bury.
REACTION TO DIVERS: Apparently relying on camouflage, remain still. Bolt only when closely approached or molested.
NOTE: Formerly classified as *Lepidopsetta bilineata*.

DISTINCTIVE FEATURES: 1. Numerous brown blotches ringed with pale yellowish brown. 2. Pointed head and snout with small mouth. 3. Top (migrating left) eye set high on head and somewhat tilted toward dorsal fin.
DESCRIPTION: Lightly mottled, spotted and blotched shades of brown. Smooth forebody, coarsely scaled toward rear. Lateral line nearly straight with long curving branch from head to near mid-dorsal fin. Oval body; margin of tail nearly straight. Relatively narrow body with arched dorsal and anal fins that form a rough diamond outline.
ABUNDANCE & DISTRIBUTION: Abundant to common southern California to Bering Sea. Also south to southern Baja.
HABITAT & BEHAVIOR: Inhabit flat sandy, silty or mud bottoms; often in vicinity of docks and jetties. Rest on bottom, frequently partially or completely covered with soft bottom material. Large adults generally below safe diving depths.
REACTION TO DIVERS: Apparently relying on camouflage, remain still. Bolt only when closely approached or molested.
NOTE: Also known as "Lemon Sole." Formerly classified in the genus *Parophrys*.

Flatfish/Bottom-Dwellers

CURLFIN SOLE
Pleuronichthys decurrens
FAMILY:
Righteye Flounder –
Pleuronectidae

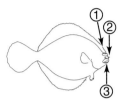

SIZE: 6-10 in.,
max. 15 in.
DEPTH: 20-1,800

ROCK SOLE
Pleuronectes bilineatus
FAMILY:
Righteye Flounder –
Pleuronectidae

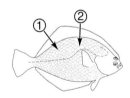

SIZE: 1-1½ ft.,
max. 2 ft.
DEPTH: 3-1,500

ENGLISH SOLE
Pleuronectes vetulus
FAMILY:
Righteye Flounder –
Pleuronectidae

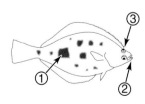

SIZE: 10-18 in.,
max. 2½ ft.
DEPTH: 0 - 1,800 ft.

Righteye Flounder

DISTINCTIVE FEATURES: 1. First five or more spinous rays of dorsal fin are elongated, with little or no webbing between. 2. Large mouth extends below lower (right) eye.
DESCRIPTION: Lightly mottled, spotted and blotched shades of brown. Lateral line nearly straight with curving branch along base of foredorsal fin. Tail rounded.
ABUNDANCE & DISTRIBUTION: Common central California to Bering Sea; uncommon to southern California.
HABITAT & BEHAVIOR: Inhabit flat sandy, silty or mud bottoms; often near docks and jetties. Rest on bottom, often partially or completely covered with soft bottom material.
REACTION TO DIVERS: Apparently relying on camouflage, remain still. Bolt only when closely approached or molested.

DISTINCTIVE FEATURES: Slender body. 1. Large noticeable scales, often with dark edges. 2. Large mouth extends below mid-eye.
DESCRIPTION: Shades of brown; often a few small white spots and pale edges on dorsal and anal fins. Single, nearly straight lateral line; tail rounded.
ABUNDANCE & DISTRIBUTION: Uncommon southern Alaska to southern California. Also south to central Baja.
HABITAT & BEHAVIOR: Inhabit flat sandy, silty or mud bottoms. Rest on bottom, often partially or completely covered with soft bottom material. Generally a deep water species that lives well below safe diving depths; at night occasionally forage in shallower depths near shore.
REACTION TO DIVERS: Apparently relying on camouflage, remain still. Bolt only when closely approached or molested.
SIMILAR SPECIES: Rex Sole, *Errex zachirus,* and Dover Sole, *Microstomus pacificus,* have small mouths and lack large noticeable scales. The Rex Sole is distinguished by a very long pectoral fin. The Dover Sole has unusually large, protruding eyes.

DISTINCTIVE FEATURES: 1. Alternating orange to yellow and gray to black bars on dorsal, anal and tail fins.
DESCRIPTION: Shades of brown to dark gray. Rough, noticeable scales cover upper side. Can be right- or left-eyed. Oval body; tail slightly rounded.
ABUNDANCE & DISTRIBUTION: Common central California to Aleutian Islands, Alaska; occasional to uncommon south to Los Angeles. Also to Japan and Korea.
HABITAT & BEHAVIOR: Inhabit flat sandy, silty or mud bottoms; often near docks and eelgrass beds. Rest on bottom, typically partially or completely covered with soft bottom material. Most common between 10-150 ft. Young often enter fresh water.
REACTION TO DIVERS: Apparently relying on camouflage, remain still. Bolt only when closely approached or molested.

Flatfish/Bottom-Dwellers

SAND SOLE
Psettichthys melanostictus
FAMILY:
Righteye Flounder –
Pleuronectidae

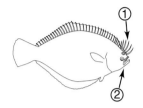

SIZE: 10-18 in.,
max. 24¾ in.
DEPTH: 3-1,000 ft.

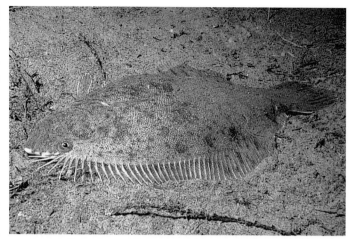

SLENDER SOLE
Eopsetta exilis
FAMILY:
Righteye Flounder –
Pleuronectidae

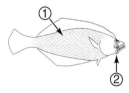

SIZE: 7-10 in.,
max. 14 in.
DEPTH: 25-1,700 ft.

STARRY FLOUNDER
Platichthys stellatus
FAMILY:
Righteye Flounder –
Pleuronectidae

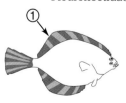

SIZE: 1-2½ ft.,
max. 3 ft.
DEPTH: 0-1,200 ft.

Righteye Flounder – Lefteye Flounder

DISTINCTIVE FEATURES: 1. Slightly concave tail. 2. Upper jaw extends to mid-eye. 3. Lateral line arches over pectoral fin.
DESCRIPTION: Shades of gray or brown, often with some mottling and spots.
ABUNDANCE & DISTRIBUTION: Uncommon to rare (within safe diving limits) southern California to Alaska. Also to Japan.
HABITAT & BEHAVIOR: Inhabit flat sandy, silty or mud bottoms. Rest on bottom, often partially or completely covered with soft bottom material. Young occasionally inhabit waters within safe diving limits; large adults much deeper.
REACTION TO DIVERS: Apparently relying on camouflage, remain still. Bolt only when closely approached or molested.

DISTINCTIVE FEATURES: 1. Tail margin arched in middle with outer edges square-cut or slightly indented. 2. Mouth large, upper jaw extends to or behind eye. 3. Lateral line arches over pectoral fin.
DESCRIPTION: Generally uniform shades of gray or brown, occasionally with some light mottling and spots. Can be right- or left-eyed.
ABUNDANCE & DISTRIBUTION: Uncommon to rare (within safe diving limits) southern California to Alaska. Also to Japan.
HABITAT & BEHAVIOR: Inhabit flat sandy, silty or mud bottoms. Rest on bottom, often partially or completely covered with soft bottom material. Young occasionally inhabit waters within safe diving limits; large adults much deeper.
REACTION TO DIVERS: Apparently relying on camouflage, remain still. Bolt only when closely approached or molested.

California Halibut
Note large mouth with upper jaw extending beyond eye.

Flatfish/Bottom-Dwellers

PACIFIC HALIBUT
Hippoglossus stenolepis
FAMILY:
Righteye Flounder –
Pleuronectidae

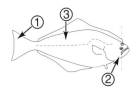

SIZE: 1½ - 3 ft.,
max. 8½ ft.
DEPTH: 25 - 3,500 ft.

CALIFORNIA HALIBUT
Paralichthys californicus
FAMILY:
Lefteye Flounder –
Bothidae

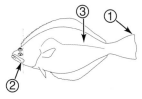

SIZE: 15 - 30 in.,
max. 5 ft.
DEPTH: 4 - 600 ft.

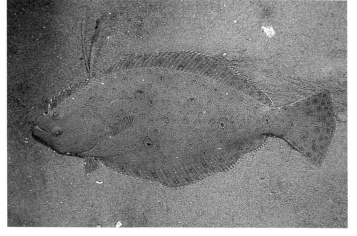

California Halibut
Even when buried in sand the size of mouth and shape of tail are distinctive of this species.

Lefteye Flounder

DISTINCTIVE FEATURES: 1. Commonly have yellow or orange spots. 2. Pectoral fin equal in length to the distance between its base and mid-eye. 3. Lateral line straight. (Of the lefteye flounders only the sanddabs have a straight lateral line.)
DESCRIPTION: Shades of brown, highly variable, often mottled or blotched with dark brown, occasionally white spots or blotches.
ABUNDANCE & DISTRIBUTION: Uncommon (abundant to common below safe diving limits) Alaska to California. Also south to southern Baja.
HABITAT & BEHAVIOR: Tend to prefer gravel bottoms, also on sand, silt or mud. Rest on bottom, often partially or completely covered with soft bottom material. Young occasionally in very shallow water; adults rarely above 60 feet with the bulk of the population below safe diving limits.
REACTION TO DIVERS: Apparently relying on camouflage, remain still. Bolt only when closely approached.
SIMILAR SPECIES: Longfin Sanddab, *C. xanthostigma*, extremely long pectoral fin which, if extended forward, would reach snout.

DISTINCTIVE FEATURES: 1. Profuse black speckles and often small blotches. 2. Pectoral fin short, fin length is less than the distance from its base to mid-eye. 3. Lateral line straight. (Sanddabs are the only lefteye flounders that have straight lateral lines.)
DESCRIPTION: Shades of light brown, often with white spots. Can change color and markings to match surroundings.
ABUNDANCE & DISTRIBUTION: Common southern Alaska to southern California. Also south to southern Baja.
HABITAT & BEHAVIOR: Inhabit sand, gravel and shell rubble flats. Rest on bottom camouflaging almost perfectly with substrate. Most common between 20-60 feet.
REACTION TO DIVERS: Apparently relying on camouflage, remain still. Bolt only when approached, swimming a short distance before resettling. Slow nonthreatening pursuit may afford a closer view. Occasionally follow divers in order to feed on small organisms disturbed by fins.

Speckled Sanddab
Color and marking variation.

Flatfish/Bottom-Dwellers

PACIFIC SANDDAB
Citharichthys sordidus
FAMILY:
Lefteye Flounder –
Bothidae

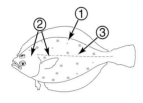

SIZE: 6-12 in.,
max. 16 in.
DEPTH: 0-1,800 ft.

SPECKLED SANDDAB
Citharichthys stigmaeus
FAMILY:
Lefteye Flounder –
Bothidae

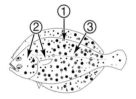

SIZE: 3-6 in.,
max. 7 in.
DEPTH: 0-1,200 ft.

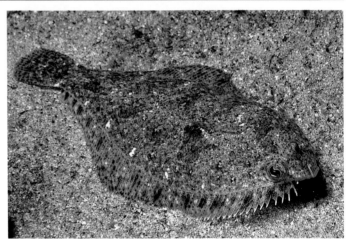

Speckled Sanddab
Color and marking variation.

IDENTIFICATION GROUP 6

Odd-Shaped Bottom-Dwellers
Poacher – Snailfish – Pipefish & Seahorse – Others

This ID Group consists of fish that normally rest on the bottom and do not have a typical fish-like shape.

FAMILY: Poacher — Agonidae
5 Species Included

Poacher (typical shape)　　　　　　　　Smooth Alligatorfish

Poachers are a small to moderate sized (averaging from four to six inches, but a few species grow to just under a foot), cold-water, bottom-dwelling family. Nearly all species live in the North Pacific. They can be recognized by their fused body plates (modified scales) that run in parallel rows down the length of their extremely tapered bodies. Sharp spines often extend from these bony plates. They have two separate dorsal fins, pairs of unusually large pectoral fins, and large, fan-like tails attached to narrow tail bases. Several species have cirri extending from under the mouth. Their pallid colors range from brown to gray, with pale undersides.

Poachers lie on sand, mud, gravel and rocky rubble bottoms. Their relatively inflexible bodies make them poor swimmers. They drag their bodies about the bottom by paddling their enlarged pectoral fins. A few species crawl by incorporating a combination of pectoral, anal and tail fin movements. Careful attention to details is required to distinguish between the many similar-appearing species.

FAMILY: Snailfish — Cyclopteridae
6 Species Included

Snailfish (typical shape)　　　　　　　　Pacific Spiny Lumpsucker

Lacking scales, snailfish have smooth flaccid skin and underlying tissue. Most have a large suction disc on the underside formed by their pectoral fins. They generally have large, smoothly rounded heads that flow into elongated bodies. Typically, sailfish have long dorsal and anal fins that nearly join the often smallish tail. Many species have a variety of color and marking patterns making them almost impossible to identify without in-hand examination. A few, however, included in this text, can be visually identified to species underwater.

Snailfish live in a wide range of habitats from intertidal zones to great depths. They use their sucking discs to hold them in place in areas of surge or current. Most family members inhabit the cooler waters in the northern hemisphere. Until recently they were classified in the family Liparididae.

FAMILY: Pipefish & Seahorse — Syngnathidae
7 Species Included

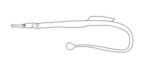

Pipefish (typical shape)

Pacific Seahorse

These strange little fish have trumpet-like snouts and small mouths. Their bodies are encased in protective bony rings made of fused scales. The heads of pipefish extend straight from their elongated bodies. Their narrow, elongated tails are capped with a small, fan-like tail fin. These rarely sighted fish move slowly about the bottom, commonly concealing themselves under marine growth and bottom debris.

Seahorses are vertically oriented, with heads cocked at right angles to their bodies, which gives them the delightful appearance of miniature horses. To maintain position they coil their long, finless tails around marine growth. Once secured, they change colors and markings to match their surroundings. When these poor swimmers move, they use their small pectoral fins to propel their listing, upright bodies through the water.

FAMILY: Others
8 Species Included

Bigeye -
Priacanthidae

Frogfish -
Antennariidae

Clingfish -
Gobiesocidae

Midshipmen/Toadfish -
Batrachoididae

Poacher

DISTINCTIVE FEATURES: Body covered with bony, armor-like plates. **1. Underslung mouth with several whisker-like cirri extending from chin. 2. Two blunt, forward-pointing spines on tip of pointed snout.** (Similar Sturgeon Poacher [next] distinguished by clumps of bushy, whisker-like cirri and sharp spines on snout.)
DESCRIPTION: Back shades of brown to gray with six or more dark bars, changing to a pale underside. Usually whitish spot on center of tail fin. Long, tapering rear body.
ABUNDANCE & DISTRIBUTION: Occasional southeastern Alaska to southern California.
HABITAT & BEHAVIOR: Inhabit sandy areas adjacent to reefs, rocky outcroppings and wall faces. Nocturnal; generally rest on bottom and occasionally move about using their large, fan-like pectoral fins; periodically stop to root in bottom material for prey. Rarely in open during day.
REACTION TO DIVERS: Apparently relying on camouflage, remain still. Move only when very closely approached or molested.

DISTINCTIVE FEATURES: Body covered with bony, armor-like plates. **1. Underslung mouth with clump of yellow, whisker-like cirri extending from tip of chin and another cluster below corners of mouth. 2. Two sharp, forward-pointing spines on tip of pointed snout.** (Similar Northern Spearnose Poacher [previous] distinguished by individual, whisker-like cirri and blunt spines on snout.)
DESCRIPTION: Back shades of brown to gray, changing to a pale underside. Enlarged head and forebody; long, tapering rear body.
ABUNDANCE & DISTRIBUTION: Occasional Aleutian Islands, Alaska, to northern California. Also Siberia, Russia.
HABITAT & BEHAVIOR: Inhabit sand, silt, gravel and other soft bottoms, and shallow eelgrass beds. Move about periodically, stopping to root in bottom material for prey. Uncommon in open during day.
REACTION TO DIVERS: Apparently relying on camouflage, remain still. Move only when very closely approached or molested.
NOTE: Formerly classified in the genus *"Agonus."*

DISTINCTIVE FEATURES: Body covered with bony, armor-like plates. **1. First dorsal fin edged with black. 2. Single spine at tip of snout. 3. One to three barbels at corners of mouth.**
DESCRIPTION: Back shades of light brown to gray with five to seven dark bars or blotches on back and sides; occasionally second dorsal fin also edged with black.
ABUNDANCE & DISTRIBUTION: Occasional Vancouver Island and southern British Columbia; rare Washington to California.
HABITAT & BEHAVIOR: Inhabit sand, silt, mud and other soft bottoms. Rest on bottom, blending with surroundings. Uncommon in open during day. Rarely encountered by divers south of Washington, where they prefer depths well below safe diving limits.
REACTION TO DIVERS: Remain still, apparently relying on camouflage. Bolt only when closely approached.

Odd-Shaped Bottom-Dwellers

NORTHERN SPEARNOSE POACHER
Agonopsis vulsa
FAMILY:
Poacher – Agonidae

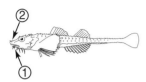

SIZE: 4-6 in., max. 8 in.
DEPTH: 30-600 ft.

STURGEON POACHER
Podothecus acipenserinus
FAMILY:
Poacher – Agonidae

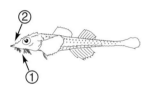

SIZE: 4-7 in. max. 1 ft.
DEPTH: 0-180 ft.

BLACKTIP POACHER
Xeneretmus latifrons
FAMILY:
Poacher – Agonidae

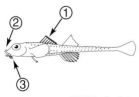

SIZE: 3-6 in.,
max. 7 1/2 in.
DEPTH: 50-1,400 ft.

135

Poacher – Bigeye

DISTINCTIVE FEATURES: Body covered with bony, armor-like plates. **1. Small, erect spine on tip of snout. 2. Two adjoining indentions on upper rear head. 3. No barbels under jaw.**
DESCRIPTION: Light shades of olive to gray with several dark blotches.
ABUNDANCE & DISTRIBUTION: Occasional southeast Alaska to Washington; rare south to southern California.
HABITAT & BEHAVIOR: Inhabit sand, silt, gravel and other soft bottoms. Masters of camouflage, lie on bottom blending with surroundings. Uncommon in open during day. Rarely encountered by divers south of Washington, where they prefer depths well below safe diving limits.
REACTION TO DIVERS: Remain still, apparently relying on camouflage. Bolt only when closely approached.

DISTINCTIVE FEATURES: Elongated body covered with smooth, bony plates. **1. Small, single dorsal fin. 2. No barbels under jaw.**
DESCRIPTION: Often uniform shades of brown to dark brown, occasionally mottled with lighter colors or with dark saddles. Two pale spots on fore-tail, one above and one below center line.
ABUNDANCE & DISTRIBUTION: Occasional Bering Sea to northern California.
HABITAT & BEHAVIOR: Inhabit sand, silt, gravel and other soft bottoms, often near rocky outcroppings or areas littered with debris from log booms. Masters of camouflage, lie on bottom blending with surroundings or resembling pieces of wood debris. Uncommon in open during day.
REACTION TO DIVERS: Remain still, apparently relying on camouflage. Bolt only when closely approached.

DISTINCTIVE FEATURES: 1. Large mouth, sharply angled upward. 2. Large eyes.
DESCRIPTION: Vary from bright red to silvery red and orange; rear dorsal, tail and anal fins often bordered in black. Compressed, saucer-shaped body.
ABUNDANCE & DISTRIBUTION: Occasional to uncommon southern and central California. Also south to Peru, including Gulf of California and Galapagos.
HABITAT & BEHAVIOR: Reclusive during day, hiding in dark protected recesses. Forage in open at night over rough, rocky bottoms and occasionally over sand for small fish, crustaceans and polychaete worms.
REACTION TO DIVERS: On night dives, mesmerized by underwater light, allowing close approach.
NOTE: Formerly classified in the genus *"Pseudopriacanthus."*

Odd-Shaped Bottom-Dwellers

PYGMY POACHER
Odontopyxis trispinosa
FAMILY:
Poacher – Agonidae

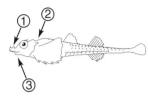

SIZE: 2-3 in., max. 4 in.
DEPTH: 15-1,200 ft.

SMOOTH ALLIGATORFISH
Anoplagonus inermis
FAMILY:
Poacher – Agonidae

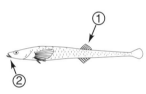

SIZE: 4-5 in., max. 6 in.
DEPTH: 15-340 ft.

POPEYE CATALUFA
Pristigenys serrula
FAMILY:
Bigeye – Priacanthidae

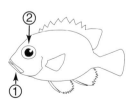

SIZE: 3-8 in., max. 1 ft.
DEPTH: 40-250 ft.

Frogfish – Snailfish

DISTINCTIVE FEATURES: Globular head and body. **1. Hand-like margins on pectoral and ventral fins. 2. Large upturned mouth.**
DESCRIPTION: Color highly variable, including shades of brown to orange or gray and uncommonly green, yellow, red or black; often spotted or blotched; frequently a large ocellated spot on rear dorsal fin and back.
ABUNDANCE & DISTRIBUTION: Rare southern California. Also south to Peru.
HABITAT & BEHAVIOR: Inhabit rocky areas. Rest on bottom, changing color to camouflage with surroundings. Fleshy protuberance between eyes can be extended and jiggled to lure prey. Poor, clumsy swimmers, often move by "walking" on fins.
REACTION TO DIVERS: Remain still, apparently relying on camouflage. Rarely move unless disturbed.

DISTINCTIVE FEATURES: Globular head and body. **1. Cone-shaped lumps cover head and body. 2. Squarish foredorsal fin widely separated from second dorsal.**
DESCRIPTION: Shades of brown to green, often with yellow or orange highlights. Wide suction cup on underside.
ABUNDANCE & DISTRIBUTION: Common to uncommon northern Washington to Aleutian Islands, Alaska. Also to Siberia, Russia.
HABITAT & BEHAVIOR: Live in wide range of habitats, including eelgrass beds to rocky areas with kelp and other algae growth; also in shallow bays and around docks. Attach to solid objects with suction cup; slow inefficient swimmers. Especially common in shallow water July to October.
REACTION TO DIVERS: Remain still, apparently relying on camouflage. Move only when disturbed.

DISTINCTIVE FEATURES: 1. Elevated foredorsal fin separated from rear by deep notch. 2. Broad, somewhat flattened head with tiny eyes.
DESCRIPTION: Nearly uniform shades of brown; occasionally line of pale spots along midbody; short, thin, dark lines radiate from eyes. Large wide mouth; dorsal and anal fin end just before tail fin.
ABUNDANCE & DISTRIBUTION: Common Washington to Alaska.
HABITAT & BEHAVIOR: Inhabit shallow rocky areas and in vicinity of docks where there is abundant algae growth. Camouflage by nestling in dense growth.
REACTION TO DIVERS: Remain still, apparently relying on camouflage. Bolt when disturbed, and disappear into foliage.
SIMILAR SPECIES: Tidepool Snailfish, *L. florae*, distinguished by size (rarely over four inches) and larger eyes; rarely below intertidal areas; southern California to Alaska. Spotted Snailfish, *L. callyodon*, distinguished by small size (rarely over five inches) and scattering of dark spots on head and body; shallow and intertidal areas; Oregon to Aleutian Islands. Slipskin Snailfish, *L. fucensis*, distinguished by light mottling and/or bands on sides and fins; southern California to Alaska.
NOTE: Formerly classified in the genus *"Polypera"* which was merged with *"Liparis."*

Odd-Shaped Bottom-Dwellers

ROUGHJAW FROGFISH
Antennarius avalonis
FAMILY:
Frogfish – Antennariidae

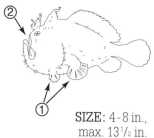

SIZE: 4 - 8 in.,
max. 13 ½ in.
DEPTH: 0 - 360 ft.

PACIFIC SPINY LUMPSUCKER
Eumicrotremus orbis
FAMILY:
Snailfish – Cyclopteridae

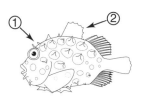

SIZE: 1 - 3 in., max. 5 in.
DEPTH: 0 - 500 ft.

LOBEFIN SNAILFISH
Liparis greeni
FAMILY:
Snailfish – Cyclopteridae

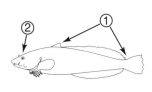

SIZE: 5 - 9 in., max. 1 ft.
DEPTH: 0 - 50 ft.

Snailfish – Clingfish

DISTINCTIVE FEATURES: 1. Dorsal, anal and tail fins continuous ending in a rounded point. 2. Foredorsal fin not elevated. 3. Head broad and somewhat flattened; with large eyes.
DESCRIPTION: Striped patterns in shades of brown; uncommonly uniform in color or with spots. Tiny mouth.
ABUNDANCE & DISTRIBUTION: Uncommon Alaska to central California. Also to Siberia, Russia.
HABITAT & BEHAVIOR: Inhabit sandy and other soft bottom areas where they generally remain hidden under scattered rocks and other shelter. Rarely venture into open during the day; often forage over bottom at night.
REACTION TO DIVERS: Remain still, apparently relying on camouflage. Bolt only when disturbed.
SIMILAR SPECIES: Slimy Snailfish, *L. mucosus*, distinguished by raised foredorsal fin and distinct tail; intertidal to 60 feet; Alaska to southern California.

DISTINCTIVE FEATURES: Skillet-shaped head and body. 1. Dark, net-like markings cover head and body. 2. Often a pale band between and below eyes.
DESCRIPTION: Mottled and blotched shades of red to brown or gray; frequently covered with minute reddish spots. Often lines or other markings radiate from eye.
ABUNDANCE & DISTRIBUTION: Occasional (but rarely observed) southern Alaska to central California; rare southern California.
HABITAT & BEHAVIOR: Cryptic; live in wide range of habitats clinging to the underside of rocks, narrow overhangs, kelp leaves and other sheltering materials.
REACTION TO DIVERS: Remain still, apparently relying on camouflage. Scurry to new refuge when disturbed.

DISTINCTIVE FEATURES: 1. Narrow head with small sucker disc on underside. 2. Single, short dorsal and anal fin.
DESCRIPTION: Emerald green to tobacco brown, changing color to match foliage; often dark or pale stripe through eye and orangish stripe on side. Slender head and body compared to typical skillet-shape of most clingfishes.
ABUNDANCE & DISTRIBUTION: Common (but rarely observed) Queen Charlotte Islands, Canada, to northern California; rare central and southern California. Also to central Baja.
HABITAT & BEHAVIOR: Inhabit kelp and eelgrass beds. When attached to plants, blend almost perfectly with background.
REACTION TO DIVERS: Remain still, apparently relying on camouflage. Bolt deep into foliage when closely approached.
SIMILAR SPECIES: Southern Clingfish, *R. dimorpha*, Slender Clingfish, *R. eigenmanni*, and Kelp Clingfish are virtually identical in appearance; close examination of captured specimen is necessary to confirm identification. Their location, however, is often useful in making an ID: Southern and Slender Clingfish are restricted to southern California while the Kelp Clingfish primarily inhabit waters to the

Odd-Shaped Bottom-Dwellers

SHOWY SNAILFISH
Liparis pulchellus
FAMILY:
Snailfish – Cyclopteridae

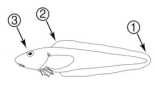

SIZE: 3 - 6 in., max. 10 in.
DEPTH: 40 - 600 ft.

NORTHERN CLINGFISH
Gobiesox maeandricus
FAMILY:
Clingfish – Gobiesocidae

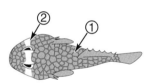

SIZE: 2 - 4 1/2 in., max. 6 1/2 in.
DEPTH: 0 - 30 ft.

KELP CLINGFISH
Rimicola muscarum
FAMILY:
Clingfish – Gobiesocidae

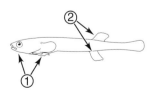

SIZE: 3/4 - 2 in., max. 2 3/4 in.
DEPTH: 1 - 60 ft.

Midshipman – Pipefish & Seahorse

DISTINCTIVE FEATURES: 1. Wide, flattened head with protruding eyes and upturned mouth. 2. Several curving rows of white spots (luminescent organs called photophores) on head; and three or four straight rows run length of body. 3. Fins unmarked.
DESCRIPTION: Shades of brown, often with purple tints and occasional blotches; sides often pale. Spinous dorsal very small with only two to four spines; second dorsal and anal fins long and even in height.
ABUNDANCE & DISTRIBUTION: Abundant to common British Columbia, Vancouver Island, Puget Sound and California; uncommon Alaska, Oregon and Pacific coast Washington.
HABITAT & BEHAVIOR: Inhabit sand and mud bottoms. Nocturnal; bury during the day, often with only upper head, eyes and mouth protruding; hover just above bottom at night feeding on small planktonic prey. Most abundant below safe diving limits.
REACTION TO DIVERS: Remain still, apparently relying on camouflage. Bolt short distance and rebury when closely approached or molested. When hovering at night, rapidly react to light by diving into silt.
SIMILAR SPECIES: Specklefin Midshipman, *P. myriaster*, distinguished by spotted pectoral and dorsal fins. Southern California.

Bay Pipefish
Brown variation with speckles.

DISTINCTIVE FEATURES: Only pipefish north of California. **1. Long trumpet-like snout. 2. Small, fan-shaped tail.**
DESCRIPTION: Shades of green to brown; belly often white, frequently speckles on head. Color and markings of pipefish are extremely variable. Habitat is often a key to identification; however, positive identification can only be made by counting body and tail rings and fin rays.
ABUNDANCE & DISTRIBUTION: Common to occasional Alaska to southern California. Also to southern Baja.
HABITAT & BEHAVIOR: Inhabit eelgrass or algae beds in shallow bays and inlets; occasionally in vicinity of docks and pilings. Slow swimmers.
REACTION TO DIVERS: Wary; retreat to cover when approached.
SIMILAR SPECIES: All three of the following species are green to brown and inhabit California waters. Barcheek Pipefish, *S. exilis*, usually dark stripe extends from lower eye; generally along coastal beaches in floats of seaweed. Barred Pipefish, *S. auliscus*, medium snout, only 14-16 body rings, may have narrow white rings; shallow bays in eelgrass beds. Kelp Pipefish, *S. californiensis*, around kelp forests, seldom in eelgrass beds.

Odd-Shaped Bottom-Dwellers

PLAINFIN MIDSHIPMAN
Porichthys notatus
FAMILY:
Toadfish – Batrachoididae

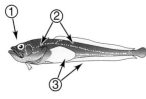

SIZE: 2-5 in., max. 15 in.
DEPTH: 0 - 1,200 ft.

Plainfin Midshipman
Typically bury in bottom material during day.

BAY PIPEFISH
Syngnathus leptorhynchus
FAMILY:
Pipefish & Seahorse – Syngnathidae

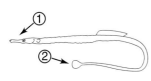

SIZE: 4-7 in., max. 13 in
DEPTH: 0 - 30 ft.

Pipefish & Seahorse

DISTINCTIVE FEATURES: 1. Long trumpet-like snout. 2. Small, fan-shaped tail. 3. Pale streak behind eye.
DESCRIPTION: Shades of brown, often with whitish blotches and spots. Color and markings of pipefish are extremely variable making them difficult to distinguish underwater. Positive identification can only be made by counting body and tail rings and fin rays. (ID of pictured specimen confirmed by counting body and tail rings.)
ABUNDANCE & DISTRIBUTION: Uncommon to rare southern California. Also south to central Baja.
HABITAT & BEHAVIOR: Generally along rocky shores in floats of seaweed; occasionally in eelgrass beds.
REACTION TO DIVERS: Wary; retreat to cover when approached.
SIMILAR SPECIES: Pugnose Pipefish, *Bryx dunckeri*, distinguished by short "pug" nose; may have narrow white rings and double row of spots down side; shallow eelgrass and seaweed beds; California.

DISTINCTIVE FEATURES: Only seahorse in eastern Pacific. **1. Horse-like head with trumpet-shaped snout and mouth. 2. Coiled tail.**
DESCRIPTION: Numerous fine, light and dark line markings radiate from eye; line markings may also decorate body and tail. Colors highly variable, including shades of gray, brown, green, yellow, gold and white. Body composed of bony rings.
ABUNDANCE & DISTRIBUTION: Rare extreme southern California. Also south to Peru, including offshore islands.
HABITAT & BEHAVIOR: Curl tail around holdfasts, changing color to camouflage with surroundings. Occasionally float free or lie on bottom.
REACTION TO DIVERS: Allow close approach and rarely move, but tuck their heads and turn away.

Odd-Shaped Bottom-Dwellers

CHOCOLATE PIPEFISH
Syngnathus euchrous
FAMILY:
Pipefish & Seahorse –
Syngnathidae

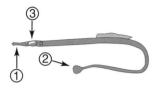

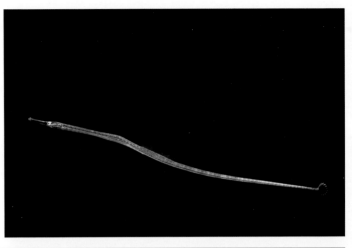

SIZE: 4 - 7 in., max. 10 in
DEPTH: 0 - 40 ft.

PACIFIC SEAHORSE
Hippocampus ingens
FAMILY:
Pipefish & Seahorse –
Syngnathidae

SIZE: 4 - 8 in., max. 1 ft.
DEPTH: 10 - 60 ft.

IDENTIFICATION GROUP 7

Odd-Shaped & Other Swimmers
Wrasse – Others

This ID Group consists of swimming fish that do not have a typical fish-like shape or do not conform to other ID Groups.

FAMILY: Wrasse — Labridae
3 Species Included

Wrasse
(typical shape)

California Sheephead
(terminal phase)

Wrasses are a large family of fish generally associated with tropical reefs; however, a few eastern Pacific species venture into the temperate waters of California. Family members are easily distinguished by protruding canine teeth, large noticeable scales and their habit of swimming with only pectoral fins. The elongated teeth are indispensable tools used for picking at and crushing the hard protective coverings of small mollusks, crustaceans and sea urchins. Wrasse are typically small (3-6 inches), but a few species, including the California Sheephead, grow much larger.

During maturation most species go through dramatic changes in color, shape and markings. These phases can include the JUVENILE PHASE (JP), INITIAL PHASE (IP) and TERMINAL PHASE (TP) which is the largest and most colorful. The initial phase includes sexually mature females, and, in some species, immature and/or mature males. Those in the terminal phase are always sexually mature males. Some wrasses are hermaphroditic and go through a sex reversal to enter the terminal phase, while others simply mature without changing sex. The three wrasse species in California waters are easily distinguished during all their phases.

FAMILY: Others
13 Species Included

Mola -
Molidae

Stickleback -
Gasterosteidae

Cardinalfish -
Apogonidae

Butterflyfish -
Chaetodontidae

Damselfish -
Pomacentridae

Ratfish/Chimaera -
Chimaeridae

Triggerfish/Leatherjacket -
Balistadae

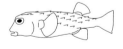

Puffer -
Tetraodontidae

Mola – Stickleback – Cardinalfish

DISTINCTIVE FEATURES: Broad oval body **1. Long dorsal and anal fins.**
DESCRIPTION: Silver to gray, gray-brown, gradating to whitish belly.
ABUNDANCE & DISTRIBUTION: Occasional southern California to British Columbia, Canada. Worldwide tropical and temperate waters.
HABITAT & BEHAVIOR: Generally oceanic, occasionally near kelp beds where they apparently go to be cleaned. Swim upright with dorsal and anal fins flapping from side to side; often tip of dorsal fin breaks surface; also lay on side "basking" at surface.
REACTION TO DIVERS: Small individuals tend to be shy and swim away when approached. Large individuals seem to be unafraid and allow slow nonthreatening approach to within arm's length.

DISTINCTIVE FEATURES: Thin, elongated body. **1. Long snout with small mouth.**
DESCRIPTION: Silver to brown, often somewhat translucent, frequently with iridescent gold, green or bluish sheen.
ABUNDANCE & DISTRIBUTION: Uncommon southern Alaska to southern California. Also to northern Baja.
HABITAT & BEHAVIOR: Form schools that inhabit kelp forests, or shallow eelgrass beds; also near pilings and often under docks. Cryptic; hover in shallows, mixing with plant growth. During late spring and early summer males build seaweed nests. A nest will be used by numerous females to deposit egg clusters which the single male fertilizes and guards.
REACTION TO DIVERS: Wary; retreat to shelter of kelp fronds or eelgrass when approached. A slow nonthreatening approach may allow close view, especially of males guarding eggs.

DISTINCTIVE FEATURES: Red to pink or orange. **1. Separated dorsal fins align with ventral and anal fins.**
DESCRIPTION: Back dark, often purplish; fin membrane translucent. Foredorsal fin dark in young.
ABUNDANCE & DISTRIBUTION: Uncommon southern California. Also Baja, including offshore islands and Gulf of California.
HABITAT & BEHAVIOR: Inhabit rocky reefs and outcroppings. Hover under ledge overhangs, in caves and other dark recesses during day. Forage in open at night.
REACTION TO DIVERS: Tend to ignore divers. Can usually be approached with slow nonthreatening movements.

Odd-Shaped & Other Swimmers

OCEAN SUNFISH
Mola mola
FAMILY:
Mola – Molidae

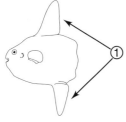

SIZE: 3-7 ft., max. 10 ft.
DEPTH: Surface

TUBE-SNOUT
Aulorhynchus flavidus
FAMILY:
Stickleback – Gasterosteidae

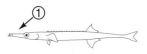

SIZE: 3-6 in., max. 7 in.
DEPTH: 4-100 ft.

GUADALUPE CARDINALFISH
Apogon guadalupensis
FAMILY:
Cardinalfish – Apogonidae

SIZE: 2-3 in., max. 5 in.
DEPTH: 20-120 ft.

Wrasse

DISTINCTIVE FEATURES: TP & IP: 1. White chin. TP: Red to orange or pink midbody; dark head and rear body. **IP:** Pink to reddish brown or reddish gray. **JP:** Red. **2. White midbody stripe. 3. Large black spots on rear dorsal and anal fins and upper base of tail.**
DESCRIPTION: TP: Over 12-14 inches in length; canine teeth protrude from front of mouth. **Older TP:** develop bulbous lump on nape. **IP:** 6-12 in length. **JP:** usually less than 6 inches in length; occasionally have a narrow, second white stripe on back.
ABUNDANCE & DISTRIBUTION: IP & JP: Occasional to uncommon. **TP:** Rare southern California. All phases uncommon to rare central California. (Formerly common, but numbers greatly reduced by overfishing, especially spearfishing of large TP.)
HABITAT & BEHAVIOR: Inhabit rocky bottoms, especially kelp beds. Generally solitary, slow-moving swimmers.
REACTION TO DIVERS: Tend to ignore divers unless closely approached or chased. A slow nonthreatening approach usually allows a close view.

**California Sheephead
Initial Phase**

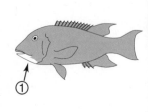

SIZE: 6 - 12 in.

DISTINCTIVE FEATURES: Yellow to orange or orange-brown. **1. Large black spot on tail base.**
DESCRIPTION: White belly. Sharp canine teeth often protrude from front of mouth.
ABUNDANCE & DISTRIBUTION: Abundant to common southern California; occasional north to northern California. Also south to central Baja.
HABITAT & BEHAVIOR: Inhabit kelp beds and rocky reefs and boulder-strewn areas surrounded by sand. Generally swim well above bottom; may be solitary, or in small groups to large schools. Cleaners that service many species of fish, including Giant Sea Bass, rays, surfperches and even Garibaldis; also, feed on wide range of invertebrates. Most common shallower than 70 feet. When frightened, dive into sand and bury; also, bury in sand to sleep at night.
REACTION TO DIVERS: Tend to ignore divers. Can often be viewed closely by moving into their direction of travel.

Odd-Shaped & Other Swimmers

CALIFORNIA SHEEPHEAD
Semicossyphus pulcher
Terminal Phase
FAMILY:
Wrasse – Labridae

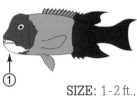

SIZE: 1 - 2 ft.,
max. 3 ft.
DEPTH: 3 - 280 ft.

**California Sheephead
Juvenile**

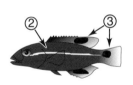

SIZE: 3 - 6 in.

SENORITA
Oxyjulis californica
FAMILY:
Wrasse – Labridae

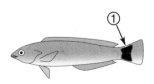

SIZE: 3 - 8 in.,
max. 10 in.
DEPTH: 0 - 320 ft.

151

Wrasse – Butterflyfish

DISTINCTIVE FEATURES: TP: 1. Dark bar behind pectoral fin. IP: 2. Dark areas on scales form spotted stripe on upper side.
DESCRIPTION: TP: Generally 12 inches or longer; green to blue-green and occasionally orange; often yellowish patch directly behind pectoral fin; may display dusky bars on back. IP: Typically between five and 12 inches in length; yellow to orange or orangish brown. JP: Generally less than five inches long; orangish brown to brown; two black spots on dorsal fin. Very small JP are bright green. Sharp canine teeth often protrude from front of mouth.
ABUNDANCE & DISTRIBUTION: Occasional southern California. Also south to southern Baja, including Gulf of California.
HABITAT & BEHAVIOR: Inhabit small rocky reefs and boulder-strewn areas mixed with sand. Generally swim well above the bottom; usually solitary or in pairs. Typically feed on small invertebrates and occasionally act as cleaners. Most common shallower than 50 feet. When frightened, dive into sand and bury; also, bury in sand to sleep at night.
REACTION TO DIVERS: Shy; rapidly dart away when approached. Occasionally, moving into their path of travel and remaining still allows a close view.

Rock Wrasse
Terminal Phase
Orange variation.

DISTINCTIVE FEATURES: 1. Black scythe-shaped bar/stripe from gill cover to upper back continuing to below base of tail.
DESCRIPTION: Yellowish silver. Black soft dorsal fin and anal fin edged with white. Yellow foredorsal fin spines.
ABUNDANCE & DISTRIBUTION: Uncommon southern California. Also south to Baja, including Gulf of California, and through Central America to Galapagos.
HABITAT & BEHAVIOR: Flit about rocky, boulder-strewn areas; often solitary. Prefer cooler, deep water.
REACTION TO DIVERS: Tend to ignore divers, but move away when approached. Best way to get a closer look is to quietly wait in a concealed position near their course of travel.

Odd-Shaped & Other Swimmers

ROCK WRASSE
Halichoeres semicinctus
Terminal Phase
FAMILY:
Wrasse – Labridae

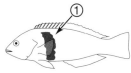

SIZE: 12 - 14 in.,
max. 15 in.
DEPTH: 0 - 80 ft.

**Rock Wrasse
Initial Phase**

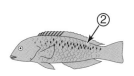

SIZE: 5 - 12 in.

SCYTHE BUTTERFLYFISH
Chaetodon falcifer
FAMILY:
Butterflyfish –
Chaetodontidae

SIZE: 3 - 4 in.,
max. 6 in.
DEPTH: 50 - 130 ft.

Damselfish

DISTINCTIVE FEATURES: Blue-gray. 1.Black spots on scales primarily scattered from midbody to tail.
DESCRIPTION: Usually blue border on dorsal, anal and tail fins.
ABUNDANCE & DISTRIBUTION: Abundant to common southern California; occasional central California. Also south to central Baja.
HABITAT & BEHAVIOR: Inhabit shallow reefs and rocky areas. Typically swim in open water, well above bottom, where they feed on plankton. Find shelter in rocky cracks, crevices and other recesses at night. Often in large aggregations or schools.
REACTION TO DIVERS: Tend to ignore divers. Can usually be approached with slow nonthreatening movements.

DISTINCTIVE FEATURES: ALL PHASES: Brilliant orange. **INTERMEDIATES:** 1. Display numerous iridescent blue spots until about six inches long. **JUVENILES:** 2. Brilliant blue spot outlined in black on mid-upper back until about two inches long.
DESCRIPTION: Thin; oval-shaped body; tail deeply notched between large rounded lobes.
JUVENILES: Numerous iridescent blue spots on head and body and borders on fins.
ABUNDANCE & DISTRIBUTION: Abundant southern California; occasional to rare Central California. Also south to southern Baja.
HABITAT & BEHAVIOR: Inhabit rocky reefs and kelp beds. Aggressively territorial, chasing away all intruders
REACTION TO DIVERS: Ignore divers or attempt to chase them away if their territory is entered.
NOTE: In California it is illegal (with severe penalties) to spear or retain Garibaldi.

**Garibaldi
Intermediate**

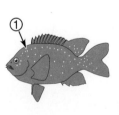

SIZE: 2 - 6 in.

Odd-Shaped & Other Swimmers

BLACKSMITH
Chromis punctipinnis
FAMILY:
Damselfish –
Pomacentridae

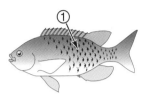

SIZE: 4-8 in.,
max. 1 ft.
DEPTH: 0-150 ft.

GARIBALDI
Hypsypops rubicundus
FAMILY:
Damselfish –
Pomacentridae

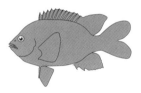

SIZE: 5-10 in.,
max. 14 in.
DEPTH: 0-95 ft.

Garibaldi Juvenile

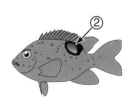

SIZE: 1-2 in.

Ratfish – Triggerfish

DISTINCTIVE FEATURES: 1. White spots over body. 2. Large, flattened snout rounded in front. 3. First dorsal fin spine tall (venomous).
DESCRIPTION: Shades of brown or gray, often with silvery sheen; commonly have iridescent tints of blue, green or gold; back darker; belly pale; fins gray. Unscaled, smooth skin. Cartilaginous fish, closely related to sharks and rays.
ABUNDANCE & DISTRIBUTION: Common northern California to British Columbia, Canada; occasional north to southeastern Alaska and south to southern California. Also south to central Baja.
HABITAT & BEHAVIOR: Inhabit sandy and muddy bottoms, and occasionally near rocky reefs. Most commonly in shallow waters, between 15-65 feet, in northern part of range, and deeper to the south. Females lay distinctive spoon-shaped egg cases [right, middle].
REACTION TO DIVERS: Wary; generally move away to maintain "safe" distance. Slow nonthreatening approach often allows a close view. At night become disoriented by diver's light and dart about erratically. Be careful to avoid foredorsal spine which is mildly toxic and can inflict a nasty wound.

Redtail Triggerfish Female
Note yellow-gold tail.

DISTINCTIVE FEATURES: Gold body with dark outlines on scales. **1. MALE:** Red tail. **FEMALE:** Yellow tail. 2. Dorsal and anal fins trimmed in brilliant red and/or yellow-gold.
DESCRIPTION: Blue line markings on head below eyes; bluish sub-border on tail.
ABUNDANCE & DISTRIBUTION: Uncommon southern California. Also south to Acapulco, Mexico, and offshore islands to Galapagos; and tropical Indo-Pacific, including Japan, Hawaii and Easter Island.
HABITAT & BEHAVIOR: Swim in open water above rocky reefs, boulder-strewn slopes and along walls.
REACTION TO DIVERS: Wary; tend to keep their distance and retreat when approached.

Odd-Shaped & Other Swimmers

SPOTTED RATFISH
Hydrolagus colliei
FAMILY:
Chimaera – Chimaeridae

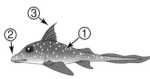

SIZE: 8-28 in.,
max. 38 in.
DEPTH: 0-3,000 ft.

Spotted Ratfish
Spoon-shaped egg case.

REDTAIL TRIGGERFISH
Xanthichthys mento
FAMILY:
Leatherjacket – Balistidae

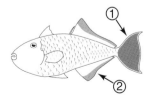

SIZE: 6-8 in.,
max. 10 in.
DEPTH: 10-80 ft.

Triggerfish – Puffer

DISTINCTIVE FEATURES: No distinctive markings.
DESCRIPTION: Drab shades of bluish gray to gray and brownish gray; belly generally lighter shade. Can pale or darken to match background. Small, fine scales on deep, rounded body.
ABUNDANCE & DISTRIBUTION: Rare California. Also south to Chile, including Galapagos.
HABITAT & BEHAVIOR: Inhabit rocky reefs, boulder-strewn slopes and adjacent areas of sand. Feed on sea urchins, small crustaceans and mollusks.
REACTION TO DIVERS: Tend to ignore divers and are occasionally somewhat curious, but keep their distance and retreat when approached. Stalking with slow nonthreatening movements may allow a close view.

DISTINCTIVE FEATURES: 1. Short, triangular, erect spines on body. 2. Prominent dark spots cover body and fins. (Similar Porcupinefish [top, next page] distinguished by long spines.)
DESCRIPTION: Bluish gray back, tan or light gray sides, gradating to white belly. Dusky band under eyes and another forward of pectoral fin.
ABUNDANCE & DISTRIBUTION: Rare southern California. Also south to Peru, and tropical and subtropical Indo-Pacific.
HABITAT & BEHAVIOR: Lurk in shaded, protective recesses in rocky reefs, boulder-strewn slopes and along walls.
REACTION TO DIVERS: Shy; retreat to protective recess when approached. Often return to peer out of entrance where they can be closely observed. Inflate if molested.
NOTE: Also commonly known as "Spotted Burrfish" and "Spottedfin Burrfish."

DISTINCTIVE FEATURES: 1. Long spines on head. 2. Small dark spots on body. No spots on fins.
DESCRIPTION: Olive to brown. Dusky band runs from eye to eye. May have dusky blotches, or bands, on back. Iris yellow; pupil has iridescent blue-green specks. Spines usually lowered, but may become erect even when the body is not inflated.
ABUNDANCE & DISTRIBUTION: Rare extreme southern California. Also south to Peru, including Galapagos; circumtropical.
HABITAT & BEHAVIOR: Lurk in shaded, protective recesses in rocky reefs, boulder-strewn slopes and along walls.
REACTION TO DIVERS: Shy; retreat to protective recess when approached. Often return to peer out of entrance where they can be closely observed. Inflate if molested.
NOTE: Also commonly known as "Barred Porcupinefish."

Odd-Shaped & Other Swimmers

FINESCALE TRIGGERFISH
Balistes polylepis
FAMILY:
Leatherjacket – Balistidae

SIZE: 1 - 2 ft.,
max. 2 ½ ft.
DEPTH: 10 - 120 ft.

PACIFIC BURRFISH
Chilomycterus affinis
FAMILY:
Puffer – Tetraodontidae

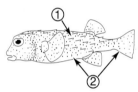

SIZE: 10 - 16 in.,
max. 20 in.
DEPTH: 10 - 90 ft.

BALLOONFISH
Diodon holocanthus
FAMILY:
Puffer – Tetraodontidae

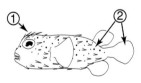

SIZE: 8 - 14 in.,
max. 20 in.
DEPTH: 10 - 50 ft.

Puffer

DISTINCTIVE FEATURES: 1. Long spines (become erect only when inflated) . 2. Small, dark spots cover entire body and fins. (Similar Pacific Burrfish [middle, previous page] distinguished by short, erect triangular spines.)
DESCRIPTION: Olive to brown or gray back, fading to whitish belly. Can pale or darken. Long spines become erect only when inflated.
ABUNDANCE & DISTRIBUTION: Rare extreme southern California. Also south to Chile, including Galapagos; circumtropical.
HABITAT & BEHAVIOR: Lurk in shaded, protective recesses in rocky reefs, boulder-strewn slopes and along walls.
REACTION TO DIVERS: Shy; retreat into protective recess when approached. Often return to peer out of entrance where they can be closely observed. Inflate if molested.

DISTINCTIVE FEATURES: 1. Pattern of concentric circles on back.
DESCRIPTION: Upper head and body shades of brown to gray, pale gray or white; underside pale. Often numerous small dark spots.
ABUNDANCE & DISTRIBUTION: Rare extreme southern California. Also south to Peru, including Galapagos and tropical Pacific.
HABITAT & BEHAVIOR: Scavenge in open water over shallow sandy areas. Bury in sand at night.
REACTION TO DIVERS: Seem to ignore divers, but tend to keep their distance.
NOTE: Also commonly known as "Concentric Puffer."

Odd-Shaped & Other Swimmers

PORCUPINEFISH
Diodon hystrix
FAMILY:
Puffer – Tetraodontidae

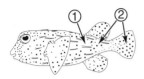

SIZE: 1-2 ft.,
max. 3 ft.
DEPTH: 10-60 ft.

BULLSEYE PUFFER
Sphoeroides annulatus
FAMILY:
Puffer – Tetraodontidae

SIZE: 8-12 in.,
max. 16 in.
DEPTH: 0-35 ft.

IDENTIFICATION GROUP 8

Silvery Swimmers
Jack – Mackerel – Surfperch – Others

This ID Group consists of fish that are silver to gray in color, and are generally unpatterned; however, many species have bluish, yellowish or greenish tints and occasional markings. Many have deeply forked tails typical of powerful open-water swimmers.

FAMILY: Jack — Carangidae
6 Species Included

Jack
(typical shape)

Almaco Jack

Jack Mackerel

Jacks are a large family of strong swimming, open-water predators that often gather in schools. Most inhabit warm seas around the world; however, a few species live in cooler regions. These powerful thin-bodied fish have small tail bases that reduce drag and deeply forked tails to facilitate speed. Their two part dorsal fins (high in the front, low in the rear) extend to the base of their tails. Silvery sides, darkish backs and large eyes are typical. As a family, jacks are easy to recognize, but it takes a sharp eye to distinguish between similar-appearing species.

FAMILY: Mackerel — Scombridae
6 Species Include

Mackerel
(typical shape)

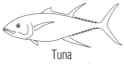
Tuna
(typical shape)

Mackerels and tunas are powerful, fast-swimming predators of the open sea. Many species typically gather in huge schools. Their long, sleek bodies and pointed snouts are somewhat cigar-shaped. Scales are small and not obvious. Their two separate dorsal fins fold into grooves. A row of small fins, called finlets, line their upper and lower rear body. Two or more keels run along each side of their narrow tail bases. The tail is widely forked. Subtle markings, slight differences in color and similar physiological characteristics make identification tricky.

FAMILY: Surfperch — Embiotocidae
11 Species Included

Surfperch (typical shape)

Kelp Perch

Black Perch

Surfperch inhabit the shallow, temperate waters of the North Pacific. These predominately silvery fish have compressed, generally oval-shaped bodies with noticeable scales. They have small, terminal mouths, a single dorsal fin and forked tails. Most have stripes, bars or other markings and color tints that make species identification possible.

FAMILY: Others
24 Species Included

Herring - Clupeidae

Anchovy - Engraulidae

Silverside - Atherinidae

Billfish - Istiophoridae

Dolphin - Coryphaenidae

Barracuda - Sphyraenidae

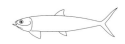
Machete/Tarpon - Elopidae

Flyingfish - Exocoetidae

Bonefish - Albulidae

Goatfish - Mullidae

Grunt - Haemulidae

Sea Chub - Kyphosidae

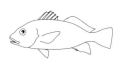
Croaker/Drum - Sciaenidae

Herring – Anchovy – Silverside

DISTINCTIVE FEATURES: 1. Row of black spots along midbody. 2. Several deep striations on gill cover. 3. Mouth extends to eye. (Similar Northern Anchovy [next] distinguished by huge mouth extending well beyond eye, and no spots along sides.)
DESCRIPTION: Silvery with noticeable scales and blue-green tints, darker on back.
ABUNDANCE & DISTRIBUTION: Occasional California; uncommon Oregon to Alaska. Also south to Baja and Gulf of California.
HABITAT & BEHAVIOR: Considered pelagic. Form huge, close-knit, polarized schools that swim in open water near shore. Often mix with similar appearing species.
REACTION TO DIVERS: Tend to ignore divers.

DISTINCTIVE FEATURES: 1. Huge, underslung mouth extends well past eye. (Similar Pacific Sardine [previous] distinguished by row of black spots along midbody and smaller mouth extending only to eye.)
DESCRIPTION: Silvery with noticeable scales and blue-green tints; back darker. Short, rounded snout.
ABUNDANCE & DISTRIBUTION: Abundant California; occasional north to Queen Charlotte Islands, British Columbia. Also south to southern Baja.
HABITAT & BEHAVIOR: Considered pelagic. Form huge, tight-knit, polarized schools that occasionally swim in open water near shore. Often near surface, especially at night.
REACTION TO DIVERS: Tend to ignore divers.

DISTINCTIVE FEATURES: 1. Anal fin begins below first dorsal fin.
DESCRIPTION: Silvery, with greenish backs and silver stripe along midbody. Short, rounded snout.
ABUNDANCE & DISTRIBUTION: Abundant California; occasional north to Queen Charlotte Islands, British Columbia. Also south to southern Baja.
HABITAT & BEHAVIOR: Tend to form aggregations in shallow waters; often in the canopy of kelp forests.
REACTION TO DIVERS: Wary; a slow nonthreatening approach may allow close view.
SIMILAR SPECIES: Jacksmelt, *A. californiensis*, distinguished by anal fin that begins behind first dorsal fin and greenish blue back. Usually over one foot in length.

Silvery Swimmers

PACIFIC SARDINE
Sardinops sagax
FAMILY:
Herring – Clupeidae

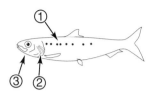

SIZE: 6 -10 in.,
max. 16 in.
DEPTH: 0 - 30 ft.

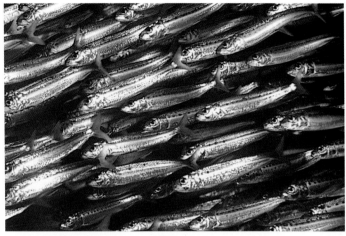

NORTHERN ANCHOVY
Engraulis mordax
FAMILY:
Anchovy – Engraulidae

SIZE: 4 - 6 in.,
max. 9 in.
DEPTH: 0 - 1,000 ft.

TOPSMELT
Atherinops affinis
FAMILY:
Silverside – Atherinidae

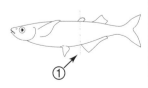

SIZE: 4 - 6 in.,
max. 9 in.
DEPTH: 0 - 1,000 ft.

165

Jack

DISTINCTIVE FEATURES: 1. Black spot on upper rear gill cover with white body streak behind.
DESCRIPTION: Silvery. Body more slender than other members of genus. Often show thin, dusky, rib-like bars.
ABUNDANCE & DISTRIBUTION: Uncommon southern California. Also south to Peru, including Gulf of California and Galapagos; and tropical Indo-Pacific.
HABITAT & BEHAVIOR: Swim rapidly in small, somewhat polarized groups or schools in open water above deep reefs, along walls and drop-offs.
REACTION TO DIVERS: Tend to ignore divers; but keep a safe distance. Apparently attracted by bubbles, occasionally make rapid, single pass and depart.

DISTINCTIVE FEATURES: 1. Yellow forked tail. 2. Dusky stripe from snout, through eye, often becomes yellowish to yellow as it continues to tail.
DESCRIPTION: Elongated, silvery body; fins often yellowish to yellow.
ABUNDANCE & DISTRIBUTION: Common to occasional southern California (mainly in spring and summer); uncommon to rare north to British Columbia, Canada. Also south through Baja, including Gulf of California to Chile; nearly worldwide in subtropical waters.
HABITAT & BEHAVIOR: Commonly cruise in open water between 10-20 feet, often under kelp floats and around offshore rigs. Frequently form small polarized schools.
REACTION TO DIVERS: Tend to ignore divers, but keep at a safe distance. Apparently attracted by bubbles, occasionally make rapid, single pass and depart.

DISTINCTIVE FEATURES: 1. Black to dusky band runs from foredorsal fin across eye to upper lip.
DESCRIPTION: Silvery to gray.
ABUNDANCE & DISTRIBUTION: Rare southern California. Also south to Peru, including Gulf of California and Galapagos; circumtropical.
HABITAT & BEHAVIOR: Swim rapidly in large, somewhat polarized, schools in open water above deep reefs, along walls and steep slopes.
REACTION TO DIVERS: Apparently attracted by bubbles; often make rapid approach, circle several times and depart.
NOTE: Formerly classified as *S. colburni* and commonly known as "Pacific Amber Jack."

Silvery Swimmers

GREEN JACK
Caranx caballus
FAMILY:
Jack – Carangidae

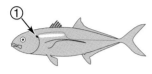

SIZE: 6 - 12 in.,
max. 15 in.
DEPTH: 10 - 180 ft.

YELLOWTAIL
Seriola lalandi
FAMILY:
Jack – Carangidae

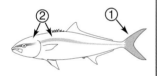
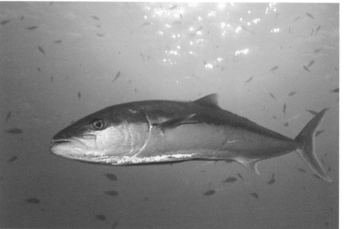

SIZE: 1 ½ - 3 ft.,
max. 5 ft.
DEPTH: 3 - 120 ft.

ALMACO JACK
Seriola rivoliana
FAMILY:
Jack – Carangidae

SIZE: 1 ½ - 2 ½ ft.,
max. 5 ft.
DEPTH: 10 - 180 ft.

Jack – Mackerel

DISTINCTIVE FEATURES: 1. Five to seven bold dark bars encircle body.
DESCRIPTION: Silvery white to gray. Long, torpedo-shaped body and forked tail.
ABUNDANCE & DISTRIBUTION: Uncommon California; rare north to Vancouver Island, Canada. Also south to Peru, including Gulf of California and Galapagos; and worldwide in tropical to warm temperate water.
HABITAT & BEHAVIOR: Accompany large fish, including sharks, rays, whales and, occasionally, ships.
REACTION TO DIVERS: Ignore divers.

DISTINCTIVE FEATURES: 1. Black spot on gill cover. 2. Midlateral line makes dramatic dip below second dorsal fin; has large distinctive scutes (bony, modified scales) along its entire length.
DESCRIPTION: Elongated, silver body; back dark in shades of silvery green to blue with iridescent highlights; silvery sheen on sides reflects a ribbed pattern.
ABUNDANCE & DISTRIBUTION: Occasional southern California to Gulf of Alaska. Also south through Baja, including Gulf of California to Galapagos.
HABITAT & BEHAVIOR: Large schools cruise in open water and along edges of kelp forests. Most common between surface and 30 feet. Young may shelter near rigs and under kelp floats; also inshore in vicinity of docks and jetties.
REACTION TO DIVERS: Tend to ignore divers.
SIMILAR SPECIES: Mexican Scad, *Decapterus scombrinus*, scutes occur only along the straight, rear lateral line. Central California to Galapagos.
NOTE: Also commonly known as "California Horse Mackerel."

DISTINCTIVE FEATURES: 1. Series of numerous, dark, slightly slanting, wavy bars on back. 2. Two dorsal fins widely separated.
DESCRIPTION: Elongated, silver body; back dark in shades of silvery green to blue, often with iridescent highlights; silvery sheen on sides reflects a ribbed pattern. About five paired dorsal and anal finlets.
ABUNDANCE & DISTRIBUTION: Common southern and central California; occasional north to Alaska. Also south to Chile; circumsubtropical and temperate.
HABITAT & BEHAVIOR: Pelagic. Swim in large polarized schools; occasionally mix with similar species. Commonly near surface when close to shore. Most abundant in summer and fall.
REACTION TO DIVERS: Tend to ignore divers.
NOTE: Also commonly known as "Pacific Mackerel."

Silvery Swimmers

PILOTFISH
Naucrates ductor
FAMILY:
Jack – Carangidae

SIZE: 6 -15 in.,
max. 2 ft.
DEPTH: 3 -100 ft.

JACK MACKEREL
Trachurus symmetricus
FAMILY:
Jack – Carangidae

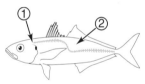

SIZE: 6 -24 in.,
max. 32 in.
DEPTH: 3 - 600 ft.

CHUB MACKEREL
Scomber japonicus
FAMILY:
Mackerel – Scombridae

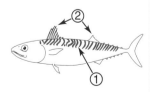

SIZE: 10 -17 in.,
max. 25 in.
DEPTH: 0-150 ft.

Mackerel – Tuna – Billfish

DISTINCTIVE FEATURES: 1. Gold spots on sides.
DESCRIPTION: Dark, silver-blue back; silvery to white sides and belly. Long, widely forked tail.
ABUNDANCE & DISTRIBUTION: Rare southern California. Also south to Peru, including Gulf of California and Galapagos.
HABITAT & BEHAVIOR: Swim in shallow water near shore, feeding on schools of small fish. Often solitary, occasionally in large schools.
REACTION TO DIVERS: Apparently attracted by bubbles; often make single, rapid pass and depart.

DISTINCTIVE FEATURES: 1. Long, yellow rear dorsal and anal fins. 2. Yellow finlets.
DESCRIPTION: Silvery, often with iridescent bluish or yellowish tints. Widely forked tail; long pectoral fin extends to base of anal fin.
ABUNDANCE & DISTRIBUTION: Occasional central and southern California. Also south to Chile; circumtropical. Commercially harvested.
HABITAT & BEHAVIOR: Inhabit clear open oceanic water. Run in large schools. Occasionally along drop-offs of offshore rocks and small islands.
REACTION TO DIVERS: Apparently attracted by bubbles; may make one or two rapid passes and depart.
SIMILAR SPECIES: Albacore or Longfin Tuna, *T. alalunga*, is distinguished by lack of yellow in fins and long pectoral fin extending beyond base of anal fin. Bigeye Tuna, *T. obesus*, distinguished by large eye and deep body. Bluefin Tuna, *T. thynnus*, distinguished by short pectoral fin and blue back and fins, no yellow.

DISTINCTIVE FEATURES: 1. Long, sword-like upper jaw (bill). 2. Long, tall, sail-like foredorsal fin is above pectoral fin (height equals or exceeds depth of body). 3. Often numerous blue bars or occasionally vertical lines ("stripes") formed by aligned spots, on back and sides (can dramatically increase or decrease intensity of these markings).
DESCRIPTION: Silvery; back dark blue to blackish; may display wide dusky bars. Midbody compressed from side to side.
ABUNDANCE & DISTRIBUTION: Common southern California. Also south to Chile and tropical Pacific.
HABITAT & BEHAVIOR: Pelagic. Near surface in open water, occasionally in vicinity of offshore islands; rarely inshore.
REACTION TO DIVERS: Shy; tend to avoid divers, but occasionally appear curious, swimming to within 15 feet. Fearless and potentially dangerous when feeding.
SIMILAR SPECIES: Blue Marlin, *Makaira nigricans*, distinguished by foredorsal fin of only moderate height. Rounded, not compressed, from side to side at midbody. Rare southern California.

Silvery Swimmers

PACIFIC SIERRA
Scomberomorus sierra
FAMILY:
Mackerel – Scombridae

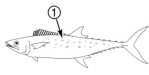

SIZE: 1 - 2 ft., max. 3 ft.
DEPTH: 0 - 40 ft.

YELLOWFIN TUNA
Thunnus albacares
FAMILY:
Mackerel – Scombridae

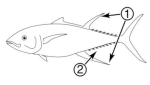

SIZE: 3 - 5 ft., max. 6 ½ ft.
DEPTH: 3 - 100 ft.

STRIPED MARLIN
Tetrapturus audax
FAMILY:
Billfish – Istiophoridae

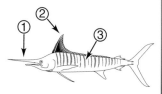

SIZE: 6 - 8 ft., max. 12 ft.
DEPTH: 0 - 40 ft.

Dolphin – Barracuda

DISTINCTIVE FEATURES: 1. Long, continuous dorsal fin extends from above eye to base of tail. **MALE (bull):** 2. Very blunt head. **FEMALE:** 3. Rounded, torpedo-shaped head.
DESCRIPTION: Brilliant silver. Males display bright yellow, yellow-green and blue iridescent spots and washes. Females display brilliant blue iridescence and washes with blue markings on head.
ABUNDANCE & DISTRIBUTION: Occasional to uncommon southern California (more abundant in warmer years); rare north to Washington. Also south to Chile; circumtropical and subtropical.
HABITAT & BEHAVIOR: Swim rapidly in open water, often under floats of kelp, seaweed and other debris. Generally in small aggregations of one or two bulls and numerous females.
REACTION TO DIVERS: Apparently curious; often make several rapid, close passes.
NOTE: Also commonly known as "Mahi mahi" and "Dorado."

DISTINCTIVE FEATURES: FEMALE: 3. Rounded, torpedo-shaped head.

DISTINCTIVE FEATURES: 1. Long, large mouth with numerous lengthy, pointed teeth and slightly jutting lower jaw. 2. Two, small, widely spaced dorsal fins.
DESCRIPTION: Elongated, silver body; back often tinted brown or iridescent blue. Thin, dark lateral line. Forked tail.
ABUNDANCE & DISTRIBUTION: Occasional southern California to Point Conception; rare north to Gulf of Alaska. Also to southern Baja, including Gulf of California.
HABITAT & BEHAVIOR: Cruise in small schools, occasionally solitary individuals in shallows, near shore waters around reefs and kelp. Have unnerving habit of working jaws, an action required to pump water through gills (not a threat).
REACTION TO DIVERS: Appear fearless, but move away when closely approached.
NOTE: Also commonly known as "California Barracuda."

Silvery Swimmers

DOLPHIN
Coryphaena hippurus
Male
FAMILY:
Dolphin - Coryphaenidae

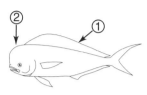

SIZE: 2 - 4 ft.,
max. 5 ¼ ft.
DEPTH: 0 - 20 ft.

Dolphin Female

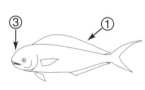

PACIFIC BARRACUDA
Sphyraena argentea
FAMILY:
Barracuda – Sphyraenidae

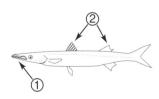

SIZE: 1 ½ - 2 ½ ft.,
max. 4 ft.
DEPTH: 3 - 120 ft.

Machete – Flyingfish

DISTINCTIVE FEATURES: 1. Long, deeply forked, darkish tail. 2. Slightly projecting lower jaw.
DESCRIPTION: Silvery with noticeable scales. Single dorsal fin. Upper jaw extends beyond rear of eye.
ABUNDANCE & DISTRIBUTION: Uncommon southern California. Also south to Peru, including Gulf of California and Galapagos.
HABITAT & BEHAVIOR: Inhabit surface waters near shore, over reefs and sandy bottoms; also in brackish lagoons and estuaries. Often school.
REACTION TO DIVERS: Apparently attracted by bubbles; often make a single, rapid pass and depart.

DISTINCTIVE FEATURES: 1. Dusky, long, wing-like pectoral fins. 2. Short snout (about equal in length to diameter of eye).
DESCRIPTION: Silvery; back dark bluish gray.
ABUNDANCE & DISTRIBUTION: Abundant to common southern California; uncommon to rare north to Oregon. Also south to southern Baja.
HABITAT & BEHAVIOR: Considered oceanic; inhabit water near surface. Rapid swimmers; when frightened, they can break the surface and glide great distances on extended pectoral fins (wings).
REACTION TO DIVERS: Shy; rapidly retreat. At night, near the surface, they may be attracted by divers' lights or boat lights, where they can be closely observed.
SIMILAR SPECIES: Blotchwing Flyingfish, *C. hubbsi*, distinguished by light band through center of pectoral fin (wing). Uncommon California.

DISTINCTIVE FEATURES: 1. Long, needle-like lower jaw. 2. Deeply forked tail, lower lobe longest.
DESCRIPTION: Silvery, often with bluish or greenish iridescent highlights; back dark. Distance between ventral fin and base of tail less than distance from ventral fin to pectoral fin.
ABUNDANCE & DISTRIBUTION: Rare southern California. Also south to Ecuador, including Galapagos.
HABITAT & BEHAVIOR: Pelagic. Inhabit open water near surface, occasionally near islands, rarely along coast. Often swim in large, rapidly moving schools.
REACTION TO DIVERS: Generally ignore divers. Apparently attracted by bubbles, occasionally make rapid, single pass and depart.
SIMILAR SPECIES: California Halfbeak, *Hyporhamphus rosae*, distinguished by tail which is only slightly indented; near shore, often in shallow bays. Silverstripe Halfbeak, *H. unifasciatus*, distinguished by tail only slightly notched, distance from eye to anal fin equal to distance from anal fin to base of tail. Ribbon Halfbeak, *Euleptorhamphus viridis*, distinguished by extremely long pectoral fins. All are rare in southern California.

Silvery Swimmers

MACHETE
Elops affinis
FAMILY:
Tarpon – Elopidae

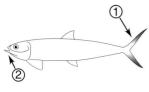

SIZE: 1 - 2 ½ ft.,
max. 3 ft.
DEPTH: 0 - 25 ft.

CALIFORNIA FLYINGFISH
Cypselurus californicus
FAMILY:
Flyingfish – Exocoetidae

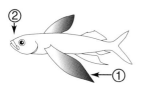

SIZE: 6 - 12 in.,
max. 19 in.
DEPTH: 0 - 25 ft.

LONGFIN HALFBEAK
Hemiramphus saltator
FAMILY:
Flyingfish– Exocoetidae

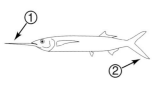
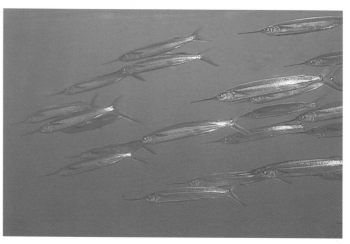

SIZE: 6 - 12 in.,
max. 1 ½ ft.,
DEPTH: 0 - 25 ft.

Bonefish – Goatfish – Surfperch

DISTINCTIVE FEATURES: 1. Short, underslung mouth that ends before eye. 2. Single dorsal fin. 3. Deeply forked tail.
DESCRIPTION: Silver. No obvious markings; darkish area at tip of snout and base of pectoral fin; may display faint bars.
ABUNDANCE & DISTRIBUTION: Uncommon southern and central California to San Francisco Bay. Also south to Peru, including Gulf of California; circumtropical.
HABITAT & BEHAVIOR: Feed over shallow flats on a rising tide, often in shallow bays and estuaries. When not feeding, may be observed on sand flats, especially around channels and inlets.
REACTION TO DIVERS: Extremely shy; difficult to approach.

DISTINCTIVE FEATURES: 1. Pair of long barbels under chin. 2. Bright yellow midbody stripe. 3. Bright yellow tail.
DESCRIPTION: White head and body, often with some bluish tinting and fine markings.
ABUNDANCE & DISTRIBUTION: Uncommon southern California. Also south to Peru, including Gulf of California and Galapagos.
HABITAT & BEHAVIOR: Inhabit sandy areas. May be solitary, but often in small groups to large aggregations, and occasionally form large polarized schools. Feed by digging in sand with barbels. At night rest on bottom, dramatically changing both color and daytime markings to reddish blotches.
REACTION TO DIVERS: Wary; tend to move away, but when busy digging in sand can often be closely approached with slow nonthreatening movements.
NOTE: Also commonly known as "Yellow-tailed Goatfish."

DISTINCTIVE FEATURES: 1. Black spots on scales form thin stripes on sides. 2. Usually two or three yellow to yellowish bars on sides.
DESCRIPTION: Thin-bodied; narrow football-shaped profile. Bright silver; usually several dusky bars on back; often dusky area on snout below nostril.
ABUNDANCE & DISTRIBUTION: Abundant southern California to British Columbia, Canada; occasional to uncommon north to southeast Alaska. Also south to central Baja.
HABITAT & BEHAVIOR: Wide range of habitats from shallow, quite backwater areas to eelgrass and kelp beds, docks and jetties, and oil rigs. Generally school during day. Most common from 0-50 feet.
REACTION TO DIVERS: Shy; usually retreat when approached. Slow nonthreatening movements may allow close view.

Silvery Swimmers

BONEFISH
Albula vulpes
FAMILY:
Bonefish - Albulidae

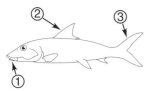

SIZE: 1 - 2 1/2 ft.,
max. 3 1/4 ft.
DEPTH: 0 - 30 ft.

MEXICAN GOATFISH
Mulloidichthys dentatus
FAMILY:
Goatfish – Mullidae

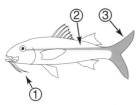

SIZE: 6 - 9 in.,
max. 1 ft.
DEPTH: 10 - 150 ft.

SHINER SURFPERCH
Cymatogaster aggregata
FAMILY:
Surfperch – Embiotocidae

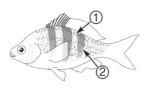

SIZE: 3 - 6 in.,
max. 7 in.
DEPTH: 0 - 480 ft.

Surfperch

DISTINCTIVE FEATURES: 1. Black line on back at base of dorsal fin. 2. Deeply forked tail, often with dusky edging.
DESCRIPTION: Thin-bodied; football-shaped profile. Silver, often with bluish iridescence or highlights. Spinous dorsal fin shorter than front part of soft dorsal.
ABUNDANCE & DISTRIBUTION: Common California; occasional Oregon to southern Vancouver Island and British Columbia. Also south to northern Baja.
HABITAT & BEHAVIOR: Inhabit shallow bays, often near docks and jetties; also offshore over sandy areas and rocky reefs, and around kelp beds. Tend to prefer calm water, but occasionally in surf and behind surflines. Swim in loose schools.
REACTION TO DIVERS: Wary, tend to keep distance. Slow nonthreatening approach may allow close view.

DISTINCTIVE FEATURES: 1. Snout pointed with upward jutting lower jaw. 2. Darkish areas on scales align to form several dusky stripes above midlateral line. 3. Head concave above eyes.
DESCRIPTION: Thin-bodied. Silver with coppery sheen to silvery copper or golden-brown, often with bluish spots; commonly one or two whitish areas below midlateral line.
ABUNDANCE & DISTRIBUTION: Abundant to common British Columbia, Canada, to northern California; uncommon to rare central and southern California. Also to central Baja.
HABITAT & BEHAVIOR: Inhabit kelp beds and occasionally under docks. Hover and mix in with kelp fronds. Generally solitary or in small groups, occasionally in large groups.
REACTION TO DIVERS: Shy; usually retreat when approached. Stalking with slow nonthreatening movements may allow close view.

Kelp Perch
Golden brown variation. Note bluish spots on lower body.

Silvery Swimmers

WHITE SEAPERCH
Phanerodon furcatus
FAMILY:
Surfperch – Embiotocidae

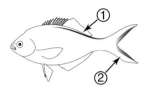

SIZE: 4-7 in.,
max. 12 ½ in.
DEPTH: 0 - 200 ft.

KELP PERCH
Brachyistius frenatus
FAMILY:
Surfperch – Embiotocidae

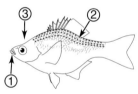

SIZE: 3 ½ - 7 in.,
max. 8 ½ in.
DEPTH: 3 - 100 ft.

Kelp Perch
Silvery copper variety. Note whitish areas below midlateral line.

179

Surfperch

DISTINCTIVE FEATURES: 1. Wide, black edge on ventral fin.

DESCRIPTION: Thin-bodied; football-shaped profile. Silver, often with bluish or greenish tints; may display dusky bars and black edging on tail and anal fins.

ABUNDANCE & DISTRIBUTION: Abundant southern and common central California; occasional northern California and Oregon; rare Washington to Vancouver Island, British Columbia. Also south to central Baja.

HABITAT & BEHAVIOR: Inhabit shallow, sandy areas; often in surf or just behind surfline; occasionally around rocky areas and shallow kelp beds, also near jetties and docks. Most common 30 feet and above. Usually in large, dense, rapidly swimming schools.

REACTION TO DIVERS: Wary, tend to keep distance. Slow nonthreatening approach may allow close view.

DISTINCTIVE FEATURES: 1. About nine dusky bars on body. 2. Narrow blue stripe along base of anal fin. 3. Patch of large scales between pectoral fin and ventral fin.

DESCRIPTION: Thin-bodied; football-shaped profile. Silvery, tinted with shades of orange to reddish brown, brown, greenish brown, green, gray or blackish; often blue to bluish specks on sides. Large lips red to orange, yellow or yellowish.

ABUNDANCE & DISTRIBUTION: Occasional central and southern California; uncommon to rare northern California. Also south to central Baja.

HABITAT & BEHAVIOR: Inhabit kelp beds. Hover one to three feet above bottom; solitary or in small groups. Rarely below 80 feet.

REACTION TO DIVERS: Shy; usually retreat when approached. Slow nonthreatening movements may allow close view.

SIMILAR SPECIES: Redtail Surfperch, *Amphistichus rhodoterus*, distinguished by body bars that offset at midlateral line, and reddish tail. Barred Surfperch, *Amphistichus argenteus*, distinguished by numerous dusky spots between bars.

DISTINCTIVE FEATURES: 1. Bars on back in shades of orange. 2. Black spot at upper corner of mouth.

DESCRIPTION: Thin-bodied; football-shaped profile except belly which is flattened. Silvery blue undercolor; scale rows below lateral line form alternating orangish and bluish silver stripes.

ABUNDANCE & DISTRIBUTION: Common southern California; occasional central California; uncommon to rare northern California. Also south to northern Baja.

HABITAT & BEHAVIOR: Inhabit sandy areas, rocky reefs, and around kelp beds. Generally solitary; occasionally gather in loose aggregations, especially in fall.

REACTION TO DIVERS: Wary; tend to keep distance. Slow nonthreatening approach may allow close view.

Silvery Swimmers

WALLEYE SURFPERCH
Hyperprosopon argenteum
FAMILY:
Surfperch – Embiotocidae

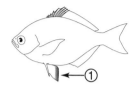

SIZE: 5 - 9 in.,
max. 1 ft.
DEPTH: 0 - 60 ft.

BLACK PERCH
Embiotoca jacksoni
FAMILY:
Surfperch – Embiotocidae

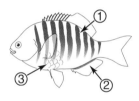

SIZE: 5 - 12 ½ in.,
max. 15 ½ in.
DEPTH: 3 - 165 ft.

RAINBOW SEAPERCH
Hypsurus caryi
FAMILY:
Surfperch – Embiotocidae

SIZE: 5 - 9 in.,
max. 1 ft.
DEPTH: 0 - 130 ft.

Surfperch

DISTINCTIVE FEATURES: 1. Numerous, narrow, iridescent blue stripes separated by wider orangish to coppery stripes.
DESCRIPTION: Thin-bodied; football-shaped profile. Silvery; back shades of gray to olive; often iridescent spots and markings on snout, head and gill cover.
ABUNDANCE & DISTRIBUTION: Common to occasional southeastern Alaska to central California and San Miguel, Santa Rosa and Santa Cruz Islands. Uncommon to rare southern California. Also south to northern Baja.
HABITAT & BEHAVIOR: Wide range of habitats from rocky reefs and kelp forests, eelgrass and leafy algae areas to sandy/rocky surf zones, and occasionally under docks. Usually solitary or in small groups; occasionally in dense schools.
REACTION TO DIVERS: Shy; usually retreat when approached. Slow nonthreatening movements may allow close view.
NOTE: Also commonly known as "Blue Surfperch."

DISTINCTIVE FEATURES: 1. Dusky to dark bar below front portion of soft dorsal fin. 2. Black spot behind corner of mouth. (Similar Rubberlip Surfperch [next] distinguished by large lips and no black spot at corner of mouth. Similar Sargo [following] distinguished by bar further forward on body, spinous dorsal taller than soft dorsal, and no spot at corner of mouth.)
DESCRIPTION: Thin-bodied; football-shaped profile. Silvery gray, occasionally brownish. Deeply forked tail. Spinous dorsal fin lower than soft.
ABUNDANCE & DISTRIBUTION: Common central California to British Columbia; occasional southeastern Alaska and southern California. Also south to northern Baja.
HABITAT & BEHAVIOR: Inhabit wide range of habitats from rocky reefs to kelp beds, under docks and around jetties and oil rigs. Often solitary or in small schools, occasionally gather in dense schools under docks and other sheltered locations. Most common between 10-65 feet.
REACTION TO DIVERS: Shy; usually retreat when approached. Slow nonthreatening movements may allow close view.

DISTINCTIVE FEATURES: 1. Large, fat lips. 2. Dusky to dark bar below front portion of soft dorsal fin.
DESCRIPTION: Thin-bodied; football-shaped profile. Dark to pale silvery gray to silvery olive or brassy brown; lips white to pink. Spinous dorsal fin lower than soft dorsal.
ABUNDANCE & DISTRIBUTION: Common southern California; occasional central California; uncommon to rare northern California. Also south to central Baja.
HABITAT & BEHAVIOR: Inhabit kelp forests, also around rocky outcroppings, jetties and piers. Often gather in schools, may mix with other surfperches. In kelp, schools often drift in mid-water just below canopy; in other environments tend to stay near bottom. Rarely in surf zones.
REACTION TO DIVERS: Wary; tend to keep distance. Slow nonthreatening approach may allow close view.

Silvery Swimmers

STRIPED SEAPERCH
Embiotoca lateralis
FAMILY:
Surfperch – Embiotocidae

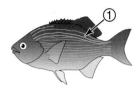

SIZE: 5 - 12 in.,
max. 15 in.
DEPTH: 10 - 70 ft.

PILE PERCH
Rhacochilus vacca
FAMILY:
Surfperch – Embiotocidae

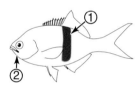

SIZE: 6 - 14 in.,
max. 17 ¼ in.
DEPTH: 0 - 260 ft.

RUBBERLIP SEAPERCH
Rhacochilus toxotes
FAMILY:
Surfperch – Embiotocidae

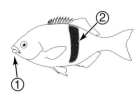

SIZE: 8 - 14 in.,
max. 18 ½ in.
DEPTH: 0 - 150 ft.

Grunt – Sea Chub

DISTINCTIVE FEATURES: 1. Dark bar below mid-spinous dorsal fin. 2. Upper gill cover edged in black and small dark area around base of pectoral fin. (Similar Rubberlip Surfperch and Pile Surfperch [previous] distinguished by dark bar further to rear and lack of black marking on gill cover and around base of pectoral fin.)
DESCRIPTION: Thin-bodied; football-shaped profile. Silver, occasionally shaded with copper; back darker. Spinous dorsal fin taller than soft. (Similar Rubberlip Surfperch and Pile Surfperch [previous] the spinous dorsal fin is shorter than soft.) **JUVENILE:** Silver with two black stripes from snout to base of tail.
ABUNDANCE & DISTRIBUTION: Common southern California; rare central California. Also south to southern Baja and northern Gulf of California.
HABITAT & BEHAVIOR: Inhabit rocky outcroppings and shallow kelp beds; often in small schools along edges of these areas, may mix with surfperches and similar fish. Also in bays, and in vicinity of docks and piers. Most common between 10-50 feet.
REACTION TO DIVERS: Wary; tend to keep distance. Slow nonthreatening approach may allow close view.

DISTINCTIVE FEATURES: 1. Yellowish bronze stripe runs from mouth to rear gill cover. 2. Dusky to yellow-bronze pinstripes run length of body
DESCRIPTION: Thin-bodied; football-shaped profile. Silvery gray body; dusky tail. Can display silvery white spots over entire body. If an imaginary line extended upward from along rear edge of the anal fin, it would pass into upper tail fin lobe.
ABUNDANCE & DISTRIBUTION: Uncommon southern California. Also south to Peru, including Gulf of California and Galapagos.
HABITAT & BEHAVIOR: Swim in somewhat polarized schools; occasionally in small groups or solitary. Most common over shallow protected areas near shore, including rocky reefs and kelp beds.
REACTION TO DIVERS: Tend to ignore divers, but move away if rapidly approached. May be possible to enter school with a very slow nonthreatening approach. Apparently attracted by bubbles; may make one or two close passes and depart.
NOTE: Also commonly known as "Striped Sea Chub."

DISTINCTIVE FEATURES: Silvery blue; darker blue back gradating to paler shades on sides and whitish belly. 1. Usually dusky spot on upper-rear gill cover.
DESCRIPTION: Football-shaped profile. May have dusky body bars.
ABUNDANCE & DISTRIBUTION: Common southern California; rare north to Vancouver Island, Canada. Also south to southern Baja, including Gulf of California.
HABITAT & BEHAVIOR: School near high profile reefs, kelp beds, oil rigs and occasionally under floating mats of vegetation.
REACTION TO DIVERS: Wary; generally keep their distance. May allow slow nonthreatening approach.

Silvery Swimmers

SARGO
Anisotremus davidsoni
FAMILY:
Grunt – Haemulidae

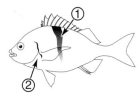

SIZE: 8 - 13 in.,
max. 17 1/4 in.
DEPTH 0 - 130 ft.

BLUE-BRONZE CHUB
Kyphosus analogus
FAMILY:
Sea Chub – Kyphosidae

SIZE: 6 - 12 in.,
max. 1 1/2 ft.
DEPTH: 3 - 40 ft.

HALFMOON
Medialuna californiensis
FAMILY:
Sea Chub – Kyphosidae

SIZE: 6 - 16 in.,
max. 19 in.
DEPTH: 0 - 130 ft.

Sea Chub – Croaker

DISTINCTIVE FEATURES: 1. One to three white spots on back. 2. Bright blue to blue-green eyes.

DESCRIPTION: Rounded, football-shaped profile. Olive-green, often shaded with blue or gray; frequently a white area on snout. Can display silvery white spot pattern over entire body.

ABUNDANCE & DISTRIBUTION: Common southern California; occasional to uncommon north to Oregon. Also south to southern Baja.

HABITAT & BEHAVIOR: Inhabit shallow rocky reefs and kelp beds; occasionally in tide pools. Solitary or in small groups. Generally near the bottom; most common between 5-60 feet.

REACTION TO DIVERS: Wary; generally move away. May allow slow nonthreatening approach.

DISTINCTIVE FEATURES: 1. Black border on upper gill cover.

DESCRIPTION: Silvery dark gray, often have brassy or purplish tints, especially on back; can lighten or darken to blend with surroundings, often light and dark scale rows align to form pinstripe pattern; fins dark. May display white spots or patches.

ABUNDANCE & DISTRIBUTION: Common southern California. Also south to southern Baja and northern Gulf of California.

HABITAT & BEHAVIOR: Inhabit rocky outcroppings, reefs and kelp beds. Lurk in sheltered areas, often in caves and crevices. Most common between 10-60 feet.

REACTION TO DIVERS: Shy; retreat to cover when approached. Slow nonthreatening movements may allow close view.

SIMILAR SPECIES: The following do not have dark gill plate border. Yellowfin Croaker, *Umbrina roncador*, silver and yellow pinstripe pattern and yellowish fins. Spotfin Croaker, *Roncador stearnsii*, silver-gray, dark patch at base of pectoral fin. Both southern California. White Croaker, *Genyonemus lineatus*, bright silver, often small dark spot at base of pectoral fin. California to Vancouver Island, Canada.

Silvery Swimmers

OPALEYE
Girella nigricans
FAMILY:
Sea Chub – Kyphosidae

SIZE: 6 - 20 in.,
max. 26 in.
DEPTH: 0 - 100 ft.

BLACK CROAKER
Cheilotrema saturnum
FAMILY:
Drum – Sciaenidae

SIZE: 6-12 in.,
max. 15 in.
DEPTH: 0 -150 ft.

IDENTIFICATION GROUP 9

Sharks & Rays

This ID Group consists of fish with skeletons composed of cartilage rather than bone, and therefore are known as cartilaginous fishes. All have small, hard scales that give them a rough, sandpapery feel. Males have claspers (long copulatory organs) at each ventral fin for internal fertilization of the female. Many species lay egg cases, called "mermaid's purses"; others keep eggs internal until hatching; and a few bear living young. Mermaid's purses are often found by divers underwater or washed up on beaches; their shapes are distinctive of species.

Sharks and rays are classified into numerous families which for the layman are difficult to distinguish and remember. Consequently, they are discussed here as two general groups rather than families.

SHARKS
15 Species Included

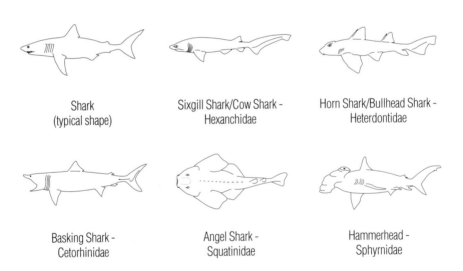

Shark
(typical shape)

Sixgill Shark/Cow Shark -
Hexanchidae

Horn Shark/Bullhead Shark -
Heterdontidae

Basking Shark -
Cetorhinidae

Angel Shark -
Squatinidae

Hammerhead -
Sphyrnidae

Most sharks have a "typical shark shape" with a heavy, long, sleek body, more-or-less pointed snout, underslung mouth and forked tail with a larger upper lobe. The members of all but two families have two dorsal fins; the first is usually the largest. They are represented by 15 families in our area.

Sharks with the "classic pointed snout," primarily requiem and mackerel sharks, are often very difficult to identify to species. Prior knowledge of their subtle differences is often required for positive identification. Many sharks, however, including hammerheads, Basking, Sixgill, Horn and Pacific Angel, have distinctive physical shapes or features that make underwater identification easy.

RAYS
13 Species Included

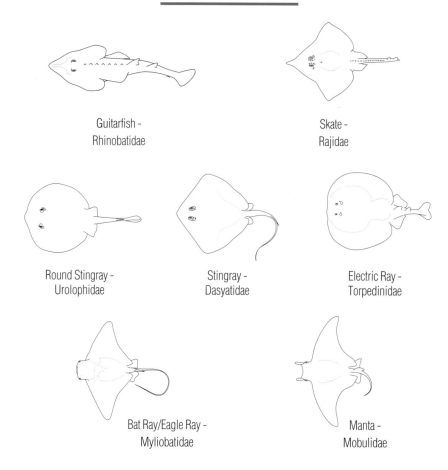

Guitarfish - Rhinobatidae

Skate - Rajidae

Round Stingray - Urolophidae

Stingray - Dasyatidae

Electric Ray - Torpedinidae

Bat Ray/Eagle Ray - Myliobatidae

Manta - Mobulidae

Rays have flattened bodies with greatly enlarged pectoral fins that are used for swimming, much like birds use their wings for flight. The pectoral fins are joined to the head and upper body, which often gives the fish a disk-like shape. Gill slits, and, in most species, the mouth, are located on the underside.

Rays are represented by eight families in our area. All are bottom dwellers, except the mantas and one species of stingray. Looking from above, the outline of most rays is distinctive of species; this, coupled with body markings, makes most rays easy to identify.

Requiem Shark – Mackerel Shark

DISTINCTIVE FEATURES: Dark blue back. **1. Long, pointed snout with underslung mouth. 2. Long pectoral fins.**
DESCRIPTION: Dark, often brilliant, blue on back changes to silver iridescent blue on sides to white belly. Slender, somewhat elongated body.
ABUNDANCE & DISTRIBUTION: Common California; occasional to uncommon Oregon to Gulf of Alaska. Also south to Chile and worldwide in subtropical and temperate waters.
HABITAT & BEHAVIOR: Inhabit open oceanic waters, commonly near surface.
REACTION TO DIVERS: Bold and unafraid, but normally not aggressive. Can be dangerous, however, especially when chum or fish blood is in water; may bite divers, apparently mistaking them as the source of odor.

DISTINCTIVE FEATURES: Large, hefty body. **1. Pointed snout. 2. Underslung mouth with readily visible large, serrated triangular teeth.** (Similar Shortfin Mako [next] easily distinguished by long, curved and pointed teeth.)
DESCRIPTION: Grayish blue to gray, blackish or brownish back gradating to lighter sides and white on belly. Often dark area at pectoral fin base. Large dorsal fin begins just behind pectoral fin; lobes of tail nearly equal in size; large tail keel.
ABUNDANCE & DISTRIBUTION: Uncommon California to Gulf of Alaska. Also worldwide distribution in temperate and subtropical waters, rare in tropical seas.
HABITAT & BEHAVIOR: Often inshore over shallow reefs and just below surf zone; most common in vicinity of sea lion and seal colonies. More common around islands; occasionally in open, offshore waters. Typically make sneak attack from below or behind. Wide ranging diet from fish to marine mammals.
REACTION TO DIVERS: Bold and unafraid. Considered very dangerous, especially if chum or fish blood is in the water. Responsible for attacks on humans, but rarely bite more than once; few encounters are fatal. Nearly all attacks on humans have been on or near the surface; it is speculated that swimmers, surfers or divers are mistaken for a sea lion or seal.

DISTINCTIVE FEATURES: Distinctly bluish with white underside. **1. Slender, conical snout. 2. Underslung mouth with readily visible long, rear curved, pointed teeth.** (Similar White Shark [previous] distinguished by large, triangular teeth with serrated edges.)
DESCRIPTION: Large dorsal fin begins just behind pectoral fin; lobes of tail nearly equal in size; large tail keel.
ABUNDANCE & DISTRIBUTION: Occasional southern California; rare north to Oregon. Also worldwide in subtropical and tropical waters.
HABITAT & BEHAVIOR: Cruise oceanic waters; rarely venture over reefs. Occasionally pass near oceanic islets and rocks. Very fast swimmers.
REACTION TO DIVERS: Bold and unafraid; often burst into view, making one or more close passes; circle and may bump. Considered quite dangerous, but usually do not attack unless chum or fish blood is in water.

Sharks & Rays

BLUE SHARK
Prionace glauca
FAMILY:
Requiem Shark –
Carcharhinidae

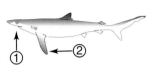

SIZE: 3-6 ft.,
max. 12 ½ ft.
DEPTH: 0-120+ ft.

WHITE SHARK
Carcharodon carcharias
FAMILY:
Mackerel Shark – Lamnidae

SIZE: 7-16 ft.,
max. 25 ft.
DEPTH: 0-4,200 ft.

SHORTFIN MAKO
Isurus oxyrinchus
FAMILY:
Mackerel Shark – Lamnidae

SIZE: 5-9 ft.,
max. 12 ½ ft.
DEPTH: 0-200 ft.

Dogfish Shark – Requiem Shark – Cow Shark

DISTINCTIVE FEATURES: 1. Sharp spine extends from and parallels front edge of each dorsal fin. 2. Anal fin absent. 3. Small underslung mouth.
DESCRIPTION: Long, flattened, pointed snout. Gray to brown back; white belly; can have white spots on back and sides.
ABUNDANCE & DISTRIBUTION: Common northern California to Aleutian Islands, Alaska; occasional central California; uncommon to rare southern California. Also, south to central Baja and worldwide in temperate waters.
HABITAT & BEHAVIOR: Often cruise over soft bottoms in shallow bays. Usually solitary, but occasionally school. Commonly feed near surface on small schooling fish.
REACTION TO DIVERS: Appear unafraid; often make surprisingly fast, close passes, usually turning away at the last moment; may bump diver. Occasionally accompany divers. Not known to actually attack.

DISTINCTIVE FEATURES: Elongated gray bodies. 1. A series of dark saddle blotches, interspaced with spots, run length of body.
DESCRIPTION: Short, bluntly rounded snout. The centers of saddles and larger spots become pale with age; occasionally specimens have only black spots. Eyes near top of head.
ABUNDANCE & DISTRIBUTION: Common northern California; occasional Oregon and central and southern California. Also south to Baja, including Gulf of California.
HABITAT & BEHAVIOR: Cruise over shallow inshore areas of sand, rocky rubble and mud flats; often in bays. In summer months often gather into schools and cruise over very shallow (ten feet or less), sandy, protected areas near the Channel Islands or California Coast. It is speculated that this behavior is related to mating.
REACTION TO DIVERS: Wary; tend to shy away when approached. Not considered a threat to divers.

DISTINCTIVE FEATURES: 1. Six gill slits. 2. Single dorsal fin near base of tail.
DESCRIPTION: Shades of gray to black or dark brown; often whitish streak down sides; belly usually lighter; greenish to bluish eyes.
ABUNDANCE & DISTRIBUTION: Occasional to uncommon southern California to British Columbia, Canada; rare to Aleutian Islands, Alaska. Also south to northern Baja.
HABITAT & BEHAVIOR: Inhabit areas with soft mud bottoms where they slowly cruise just above the bottom as they scavenge for bottom dwelling fish. Often near the edges of deep rocky reefs; occasionally young in shallow bays, inlets and inshore areas; large adults generally below safe diving limits. Very young may rest on bottom. Bear live young.
REACTION TO DIVERS: Generally docile and tend to ignore divers. Approach with caution, however; annoying any shark can provoke an attack.
NOTE: Also commonly known as "Mud Shark."
SIMILAR SPECIES: Sevengill Shark, *Notorynchus cepedianus*, has same body profile, but can be distinguished by seven gill slits and scattered black spots on body.

Sharks & Rays

SPINY DOGFISH
Squalus acanthias
FAMILY:
Dogfish Shark – Squalidae

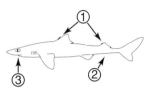

SIZE: 2-4 ft., max. 5 ft.
DEPTH: 10-120+ ft.

LEOPARD SHARK
Triakis semifasciata
FAMILY:
Requiem Shark – Carcharhinidae

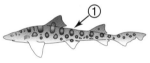

SIZE: 2-5 ½ ft., max. 7 ft.
DEPTH: 0-300 ft.

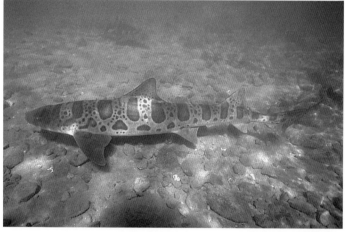

SIXGILL SHARK
Hexanchus griseus
FAMILY:
Cow Shark – Hexanchidae

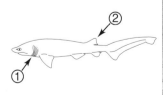

SIZE: 5-10 ft., max. 15 ½ ft.
DEPTH: 10-300+ ft.

193

Bullhead Shark – Cat Shark

DISTINCTIVE FEATURES: 1. Prominent spine at front of each of the two widely spaced dorsal fins. (Similar Swell Shark [next] distinguished by lack of spines and closely spaced dorsal fins toward rear of body.) **2. Scattered black spots over entire body (juveniles may have white spots). 3. Prominent ridges above eyes.**
DESCRIPTION: Shades of tan to brown, occasionally gray. Blunt snout.
ABUNDANCE & DISTRIBUTION: Occasional southern California; rare central California. Also south to Baja, including Gulf of California.
HABITAT & BEHAVIOR: Inhabit rocky reefs and kelp beds. Reclusive during day, hiding in caves, recesses and under algae fronds; actively forage at night.
REACTION TO DIVERS: Generally docile, tend to ignore divers. Can usually be approached with slow nonthreatening movements.

Swell Shark
Young emerging from egg case. Cases are commonly known as "mermaids' purses."

DISTINCTIVE FEATURES: 1. Dark brown blotches and spots over body.
DESCRIPTION: Light reddish brown to tan. Flattened head, two dorsal fins located well back on body.
ABUNDANCE & DISTRIBUTION: Common southern California; occasional central California. Also south to Chile.
HABITAT & BEHAVIOR: Inhabit rocky reefs, kelp beds and sand flats. Normally rest on bottom in caves and crevices or under ledge overhangs; they occasionally swim sluggishly about, near the bottom. Most common between 15-120 feet. When frightened, can greatly inflate body with water.
REACTION TO DIVERS: Docile; usually do not react unless harassed.
NOTE: Also commonly known as "Balloon Shark" or "Puffer Shark."

Sharks & Rays

HORN SHARK
Heterodontus francisci
FAMILY:
Bullhead Shark –
Heterodontidae

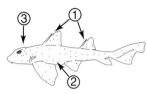

SIZE: 1½ - 2½ ft.,
max. 3¼ ft.
DEPTH: 0-500 ft.

Horn Shark
Young emerging from distinctively spiraled egg case. Cases are commonly known as "mermaids' purses."

SWELL SHARK
Cephaloscyllium ventriosum
FAMILY:
Cat Shark – Scyliorhinidae

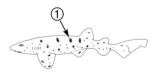

SIZE: 1 - 2½ ft.,
max. 3¼ ft.
DEPTH: 15 -1,500 ft.

Cat Shark – Basking Shark – Hammerhead

DISTINCTIVE FEATURES: Small. 1. Broad, rounded, flattened snout. 2. Two dorsal fins set well back on body near base of tail.
DESCRIPTION: Shades of brown to nearly black.
ABUNDANCE & DISTRIBUTION: Uncommon northern California to British Columbia, Canada; rare southern and central California and to southeastern Alaska. Also south to northern Baja.
HABITAT & BEHAVIOR: Inhabit sand, mud and gravel bottoms. Rest on bottom, occasionally swim sluggishly about. Generally inhabit depths considerably deeper than safe diving limits. Females lay rectangular egg cases with tendrils at each corner used to secure the cases to gorgonians, debris, etc.
REACTION TO DIVERS: Docile; tend to ignore divers.

DISTINCTIVE FEATURES: 1. Enormous white mouth. 2. Long gill slits, almost meet on underside.
DESCRIPTION: Shades of brown to gray or black, often with whitish patches. Pointed snout, sickle-shaped tail.
ABUNDANCE & DISTRIBUTION: Occasional southern California to Gulf of Alaska. Also south to Baja, Gulf of California and Chile; worldwide in subtropical and temperate waters.
HABITAT & BEHAVIOR: Cruise in both coastal and pelagic waters, often in groups, at or near surface, with wide-open mouths straining plankton from water. Close mouths approximately every 45 seconds to swallow accumulated food. Seasonally migrate to deep water. Pups at birth are five feet or slightly larger.
REACTION TO DIVERS: Tend to ignore divers. Due to large size may be dangerous if harassed.

DISTINCTIVE FEATURES: 1. Front edge of "hammer" slightly rounded and scalloped. 2. Inside tip of pectoral fin dark. 3. Rear edge of ventral fin straight.
DESCRIPTION: Gray with pale underside.
ABUNDANCE & DISTRIBUTION: Uncommon southern California. Also Baja, including Gulf of California to Ecuador and Galapagos Islands, and worldwide in tropical waters.
HABITAT & BEHAVIOR: Considered oceanic, although occasionally around islands, islets and rocky underwater pinnacles where they cruise over shallow reefs, sandy areas and along walls. May gather into small groups to huge schools numbering over 100 individuals.
REACTION TO DIVERS: Wary; generally move away when approached, but occasionally make close passes. Considered dangerous, especially in the vicinity of spearfishing.
SIMILAR SPECIES: Great Hammerhead, *S. mokarran*, distinguished by a relatively flat front edge of "hammer"; inside tip of pectoral fin not dark; rear edge of ventral fin curved. Smooth Hammerhead, *S. zygaena*, distinguished by a slightly rounded front edge of "hammer" that is smooth and not scalloped.

Sharks & Rays

BROWN CAT SHARK
Apristurus brunneus
FAMILY:
Cat Shark – Scyliorhinidae

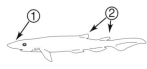

SIZE: 8 - 18 in.,
max. 2 ¼ ft.
DEPTH: 100 - 4,000 ft.

BASKING SHARK
Cetorhinus maximus
FAMILY:
Basking Shark –
Cetorhinidae

SIZE: 20 - 25 ft.,
max. 40 ft.
DEPTH: 0 - 120+ ft.

SCALLOPED HAMMERHEAD
Sphyrna lewini
FAMILY:
Hammerhead – Sphyrnidae

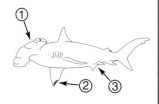

SIZE: 5 - 9 ft., max. 14 ft.
DEPTH: 20 - 160 ft.

Angel Shark – Guitarfish

DISTINCTIVE FEATURES: Similar in appearance to rays. **1. Blunt head with mouth in front.** (Similar appearing guitarfishes, skates and rays [following] have pointed heads with mouth on underside.)
DESCRIPTION: Flattened head and forebody with large pectoral fins. Generally white with tints of red, brown or gray; occasionally dark brown to nearly black. Spotted and blotched in shades of brown to black. Rear body and base of tail tubular; two dorsal fins on rear body near base of tail.
ABUNDANCE & DISTRIBUTION: Occasional California; uncommon to rare Oregon to southern Alaska. Also Baja, including Gulf of California, Peru and Chile. A vanishing species due to overfishing.
HABITAT & BEHAVIOR: Inhabit sand and other soft bottom, often near rocky reefs and kelp beds. Rest on bottom, may partially bury in bottom material.
REACTION TO DIVERS: Tend to ignore divers; however, when resting on or buried in bottom, may bite if touched, especially if grabbed by the tail.

DISTINCTIVE FEATURES: 1. Long, V-shaped head. 2. Single row of spines down back and tail.
DESCRIPTION: Brown to yellow-brown to gray, may be lightly blotched and mottled. Flattened, relatively narrow disc; long, thick tail base with two prominent dorsal fins and large tail fin.
ABUNDANCE & DISTRIBUTION: Common southern California; occasional north to Monterey; rare to San Francisco. Also Baja, including Gulf of California. Can be locally abundant.
HABITAT & BEHAVIOR: Rest on sand or mud bottoms, partially or completely buried, with only eyes protruding. Often under surf line where they regularly feed.
REACTION TO DIVERS: Tend to ignore divers; bolt only when approached within about five feet.

DISTINCTIVE FEATURES: 1. Dusky blotches form bars across back and tail base. 2. Single row of short spines run down back and tail.
DESCRIPTION: Usually shades of brown, occasionally gray; often have pale spots on back. Flattened, relatively narrow disc; long, thick tail base with two prominent dorsal fins and large tail fin.
ABUNDANCE & DISTRIBUTION: Rare southern California. Also Baja, including Gulf of California to Panama.
HABITAT & BEHAVIOR: Inhabit rocky reefs and gravel-strewn areas. Commonly rest in crevices and other recesses; rarely bury in sand.
REACTION TO DIVERS: Tend to ignore divers, moving only when closely approached or molested.

Sharks & Rays

ANGEL SHARK
Squatina californica
FAMILY:
Angel Shark – Squatinidae

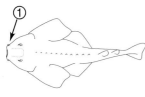

SIZE: 2 - 4 ft., max. 5 ft.
DEPTH: 10 - 600 ft.

SHOVELNOSE GUITARFISH
Rhinobatos productus
FAMILY:
Guitarfish – Rhinobatidae

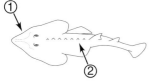

SIZE: 2 - 3 ½ ft., max. 5 ½ ft.
DEPTH: 0 - 80 ft.

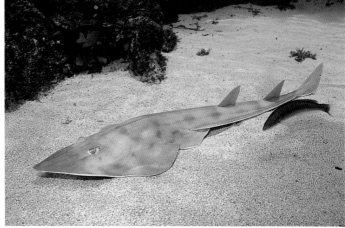

BANDED GUITARFISH
Zapteryx exasperata
FAMILY:
Guitarfish – Rhinobatidae

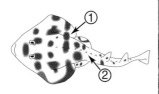

SIZE: 1 ½ - 2 ½ ft., max. 3 ft.
DEPTH: 0 - 70 ft.

Guitarfish – Skate

DISTINCTIVE FEATURES: 1. Three parallel rows of large, curved spines run down back and base of tail to just past first dorsal fin.
DESCRIPTION: Shades of brown; white underside. Disc-shaped flattened body slightly wider than long; long, thick tail base with two prominent dorsal fins just before fan-shaped tail fin.
ABUNDANCE & DISTRIBUTION: Common southern California; occasional to uncommon central California; rare northern California. Also south to Baja.
HABITAT & BEHAVIOR: Rest on sand and mud bottoms where they often partially or completely bury. Most common in water less than 25 feet.
REACTION TO DIVERS: Unafraid; do not move unless closely approached or molested.
NOTE: Formerly classified in family Platyrhinidae.

DISTINCTIVE FEATURES: 1. Long pointed snout. 2. Black spot with pale center and border on each pectoral fin. (Similar Big Skate [previous] distinguished by black spot without pale center and less pronounced snout.) 3. Row of spines above each eye and another down rear body and tail.
DESCRIPTION: Shades of brown; often have pale and dark spots on back. Leading edge of flattened body (pectoral fins) slightly concave with pointed tips; long narrow tail.
ABUNDANCE & DISTRIBUTION: Uncommon southern Alaska to southern California.
HABITAT & BEHAVIOR: Rest on sand flats where they are often partially or completely buried with only eyes protruding. Females lay rectangular, flattened egg cases with hook-like extensions at each corner.
REACTION TO DIVERS: Apparently relying on camouflage, seldom move unless closely approached or molested.
SIMILAR SPECIES: California Skate, *R. inornata*, distinguished by rounded pectoral fin tips; may have pair of ocellated spots on pectoral fins. Inhabit shallow, inshore waters and bays; rarely deeper than 40 feet. California to Washington.

DISTINCTIVE FEATURES: 1. V-shaped snout. 2. Large black spot with pale border and occasionally narrow black outer-ring near center of each pectoral fin. (Similar Longnose Skate [next] distinguished by black spot with pale center and more pronounced snout.) 3. Row of spines down back and tail to first dorsal fin.
DESCRIPTION: Shades of brown or gray to nearly black; often have white spots and mottling. Leading edge of flattened body (pectoral fins) slightly concave with pointed tips; long narrow tail.
ABUNDANCE & DISTRIBUTION: Common northern California to Bering Sea; occasional central California; rare southern California. Also northern Baja.
HABITAT & BEHAVIOR: Inhabit sand flats. Rest on bottom, often partially or completely buried with only eyes protruding.
REACTION TO DIVERS: Apparently relying on camouflage, seldom move unless closely approached or molested.
SIMILAR SPECIES: Starry Skate; *R. stellulata*, distinguished by rounded pectoral fin tips and numerous tiny spines on body; often numerous small spots. Inhabit soft bottoms, rarely shallower than 80 feet. California to Alaska.

Sharks & Rays

THORNBACK
Platyrhinoidis triseriata
FAMILY:
Guitarfish – Rhinobatidae

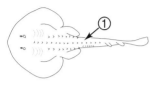

SIZE: 1-2 ft., max. 3 ft.
DEPTH: 0-150 ft.

LONGNOSE SKATE
Raja rhina
FAMILY:
Skate – Rajidae

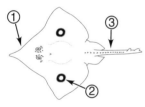

SIZE: 2-3 ft., max. 4 ft.
DEPTH: 60-2,250 ft.

BIG SKATE
Raja binoculata
FAMILY:
Skate – Rajidae

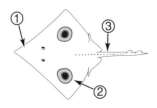

SIZE: 3-6 ft., max. 8 ft.
DEPTH: 10-350 ft.

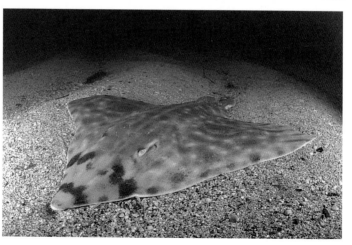

Stingray – Round Stingray

DISTINCTIVE FEATURES: 1. Pointed snout. 2. Pointed "wing" tips form diamond shape. 3. Tail thick and cylindrical to venom-injecting barb; then changes, becoming thinner and whip-like.
DESCRIPTION: Olive-brown to brown or gray back; white underside. No markings.
ABUNDANCE & DISTRIBUTION: Occasional southern California; rare north to British Columbia, Canada. Also Baja, Gulf of California and south to Peru, including Galapagos Islands.
HABITAT & BEHAVIOR: Inhabit sand, mud and gravel/rubble bottoms; often in sand around kelp beds. Rest on bottom, often partially or completely bury with only eyes protruding; when moving, glide over bottom using a wave-like body motion. Dig in sand to feed.
REACTION TO DIVERS: Tend to ignore divers, bolting only when closely approached.

DISTINCTIVE FEATURES: Only stingray to commonly swim in open water well off bottom. **1. Rounded forebody.** (Similar Diamond Stingray [previous] distinguished by bottom-dwelling behavior and angular forebody and pointed snout.) **2. Long, whip-like tail.**
DESCRIPTION: Shades of brown to dark gray, often with purplish tint; underside shades of gray. (Similar Diamond Stingray has white underside). No markings. Long venom-injecting barb at about mid-tail.
ABUNDANCE & DISTRIBUTION: Rare west coast United States. Also south to Ecuador, including Gulf of California and Galapagos, and worldwide in warm temperate and tropical waters.
HABITAT & BEHAVIOR: Considered pelagic. Swim in open water, occasionally entering deep bays and sounds.
REACTION TO DIVERS: Tend to ignore divers, moving away only when closely approached.

DISTINCTIVE FEATURES: 1. **Circular body.** 2. **Short, thick tail with broad, rounded fin.** (Similar Pelagic and Diamond Stingrays [previous] distinguished by diamond-shaped bodies and long whip-like tails without fin.)
DESCRIPTION: Usually shades of brown to occasionally gray, rarely black; often have pale spots on back. Venom-injecting barb at about mid-tail.
ABUNDANCE & DISTRIBUTION: Common southern California; uncommon north to central California; rare to northern California. Also south to Panama, including Gulf of California.
HABITAT & BEHAVIOR: Inhabit sand and mud bottoms. Young to about 7 inches remain in shallow water, rarely deeper than 15 feet; with maturity move to deeper water, most common between 35-55 feet. Rest on bottom, often partially or completely buried with only eyes protruding; when moving, glide over bottom using a wave-like body motion. Dig in sand to feed.
REACTION TO DIVERS: Tend to ignore divers; bolt when closely approached.

Sharks & Rays

DIAMOND STINGRAY
Dasyatis dipterura
FAMILY:
Stingray – Dasyatidae

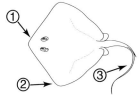

SIZE: 3 1/2 - 6 ft.
(not including tail)
DEPTH: 10 - 200 ft.

PELAGIC STINGRAY
Dasyatis violacea
FAMILY:
Stingray – Dasyatidae

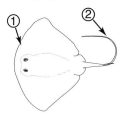

SIZE: 3 - 5 1/2 ft.
(not including tail)
DEPTH: 80 - 750 ft.

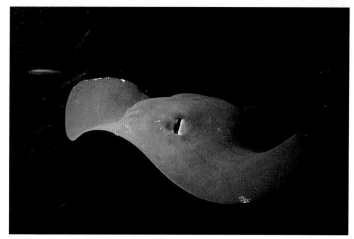

ROUND STINGRAY
Urolophus halleri
FAMILY:
Round Stingray – Urolophidae

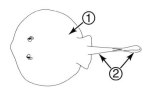

SIZE: 8 - 16 in.,
max. 22 in.
DEPTH: 0 - 70 ft.

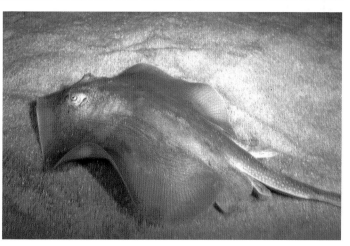

Electric Ray – Bat Ray – Manta

DISTINCTIVE FEATURES: Thick round body. 1. Short tail base with large fin.
DESCRIPTION: Shades of brown to gray, may have bluish tint; usually have scattering of black spots. Body smooth with no rows of spines on back or tail.
ABUNDANCE & DISTRIBUTION: Common southern California; occasional north to northern British Columbia, Canada. Also south to central Baja.
HABITAT & BEHAVIOR: Inhabit wide range of habitats from sand and mud flats to rocky areas and kelp forests. Normally hover or slowly swim about; occasionally rest on bottom, partially or completely buried with only eyes protruding. Forage above bottom at night. Stun prey (fish) with electric shock.
REACTION TO DIVERS: Unafraid; may swim straight at diver. If touched, can deliver a painful electric shock!

DISTINCTIVE FEATURES: 1. Large bulbous, blunt head. 2. Pectoral fins long, nearly forming equilateral triangles. 3. Whip-like tail.
DESCRIPTION: Shades of gray to brown or olive to nearly black; white underside. Long venom injecting barb at base of tail.
ABUNDANCE & DISTRIBUTION: Common southern to northern California; uncommon Oregon. Also south to Baja and Gulf of California.
HABITAT & BEHAVIOR: Inhabit wide range of habitats from sand and mud flats to kelp beds. Swim alone or in schools or rest on bottom. More active at night, swimming in midwater to the surface.
REACTION TO DIVERS: Unafraid; usually allow a slow, nonthreatening approach, but may bolt when a diver comes within five to eight feet. May bump divers at night.

DISTINCTIVE FEATURES: 1. Large mouth on leading edge of head with conspicuous, movable, scoop-like fins (pectoral) on either side.
DESCRIPTION: Black to dark gray back, often with whitish patches on shoulder and occasionally other areas. White underside, often with grayish or black areas and blotches.
ABUNDANCE & DISTRIBUTION: Uncommon southern California. Also Baja, Gulf of California and south to Peru, including Galapagos Islands. Circumtropical.
HABITAT & BEHAVIOR: Considered oceanic. Occasionally cruise along walls and over reefs. Usually solitary, but occasionally in small groups.
REACTION TO DIVERS: Tend to ignore divers unless closely approached, which may cause them to move away.

Sharks & Rays

PACIFIC ELECTRIC RAY
Torpedo californica
FAMILY:
Electric Ray – Torpedinidae

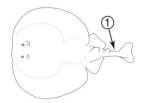

SIZE: 2-3½ ft.,
max. 4½ ft.
DEPTH: 10-1,000 ft.

BAT RAY
Myliobatis californica
FAMILY:
Eagle Ray – Myliobatidae

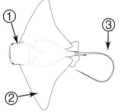

SIZE: 2-5 ft.,
max. 6 ft. (wide)
DEPTH: 0-150 ft.

MANTA
Manta birostris
FAMILY:
Manta – Mobulidae

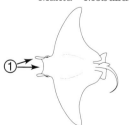

SIZE: Wing span 6-12 ft.,
max. 22 ft.
DEPTH: 0-80 ft.

COMMON NAME INDEX

A
Alligatorfish, Smooth, 137
Anchovy, Northern, 165

B
Balloonfish, 159
Barracuda, Pacific, 173
Bass, Kelp, 25
Blacksmith, 155
Blenny, Deepwater, 103
   Bay, 114
   Mussel, 114
   Rockpool, 115
Bocaccio, 41
Bonefish, 177
Brotula, Purple, 115
   Red, 113
Burrfish, Pacific, 159
Butterflyfish, Scythe, 153

C
Cabezon, 57
Cardinalfish, Guadalupe, 149
Catalufa, Popeye, 137
Chub, Blue-bronze, 185
Clingfish, Kelp, 141
   Northern, 141
   Slender, 140
   Southern, 140
Cockscomb, High, 85
   Slender, 85
Cod, Pacific, 111
Combfish, Longspine, 111
Conger, Catalina, 79
Croaker, Black, 187
   Spotfin, 186
   White, 186
   Yellowfin, 186
Cusk-eel, Spotted, 79

D
Dogfish, Spiny, 193
Dolphin, 173

E
Eelpout, Black, 88
   Blackbelly, 89
   Shortfin, 88
Electric Ray, Pacific, 205

F
Flounder, Starry, 127

Flyingfish, Blotchwing, 174
   California, 175
Fringehead, Onespot, 97
   Sarcastic, 97
   Yellowfin, 99
Frogfish, Roughjaw, 139

G
Garibaldi, 155
Goatfish, Mexican, 177
Goby, Bay, 119
   Blackeye, 117
   Bluebanded, 119
   Zebra, 119
Greenling, Kelp, 107
   Painted, 105
   Rock, 107
   Whitespotted, 109
Grouper, Broomtail, 23
   Gulf, 25
Guitarfish, Banded, 199
   Shovelnose, 199
Gunnel, Crescent, 91
   Kelp, 88
   Longfin, 91
   Penpoint, 89
   Rockweed, 89
   Saddleback, 93

H
Hagfish, Pacific, 93
Hake, Pacific, 110
Halfbeak, California, 174
   Longfin, 175
   Ribbon, 174
   Silverstripe, 174
Halfmoon, 185
Halibut, California, 129
   Pacific, 129
Hammerhead, Great, 196
   Scalloped, 197
   Smooth, 196

I
Irish Lord, Brown, 60
   Red, 61

J
Jack, Almaco, 167
   Green, 167
Jacksmelt, 164

K

Kelpfish, Crevice, 101
    Giant, 101
    Island, 103
    Spotted, 105
    Striped, 100

L

Lingcod, 109
Lizardfish, California, 113
Lumpsucker, Pacific Spiny, 139

M

Machete, 175
Mackerel, Chub, 169
    Jack, 169
Mako, Shortfin, 191
Manta, 205
Marlin, Blue, 170
    Striped, 171
Midshipman, Plainfin, 143
    Specklefin, 142
Moray, California, 79

O

Opaleye, 187

P

Perch, Black, 181
    Kelp, 179
    Pile, 183
Pikeblenny, Orangethroat, 99
Pilotfish, 169
Pipefish, Barcheek, 142
    Barred, 142
    Bay, 143
    Chocolate, 145
    Kelp, 142
    Pugnose, 144
Poacher, Blacktip, 135
    Northern Spearnose, 135
    Pygmy, 137
    Sturgeon, 135
Pollock, Walleye, 111
Porcupinefish, 161
Prickleback, Black, 83
    Monkeyface, 83
    Ribbon, 82
    Rock, 82
    Snake, 87
Puffer, Bullseye, 161

Q

Quillfish, 93

R

Ratfish, Spotted, 157
Ray, Bat, 205
    Pacific Electric, 205
Rockfish, Black, 43
    Black-and-yellow, 27
    Blue, 43
    Brown, 37
    Calico, 29
    Canary, 35
    China, 27
    Copper, 31
    Dusky, 43
    Flag, 49
    Gopher, 29
    Grass, 37
    Greenstriped, 49
    Honeycomb, 45
    Kelp, 37
    Olive, 39
    Puget Sound, 35
    Quillback, 29
    Redstripe, 33
    Rosy, 45
    Silvergray, 41
    Squarespot, 35
    Starry, 45
    Tiger, 47
    Vermilion, 33
    Widow, 39
    Yelloweye, 31
    Yellowtail, 39
Ronquil, Northern, 117
    Stripedfin, 117

S

Sand Bass, Barred, 27
    Spotted, 25
Sanddab, Longfin, 130
    Pacific, 131
    Speckled, 131
Sardine, Pacific, 165
Sargo, 185
Scad, Mexican, 168
Scorpionfish, California, 53
    Rainbow, 55
    Stone, 53

Sculpin, Bald, 70
  Buffalo, 59
  Coralline, 63
  Great, 55
  Grunt, 75
  Lavender, 73
  Longfin, 71
  Manacled, 73
  Mosshead, 70
  Northern, 67
  Pacific Staghorn, 57
  Padded, 61
  Roughback, 67
  Roughspine, 65
  Sailfin, 75
  Scalyhead, 63
  Silverspotted, 75
  Smoothhead, 60
  Snubnose, 65
  Soft, 73
  Spinyhead, 59
  Spotfin, 69
  Threadfin, 69
  Woolly, 71
Sea Bass, Giant, 23
Seahorse, Pacific, 145
Seaperch, Rainbow, 181
  Rubberlip, 183
  Striped, 183
  White, 179
Senorita, 151
Shark, Angel, 199
  Basking, 197
  Blue, 191
  Brown Cat, 197
  Horn, 195
  Leopard, 193
  Sevengill, 192
  Sixgill, 193
  Swell, 195
  White, 191
Sheephead, California, 151
Sierra, Pacific, 171
Skate, Big, 201
  California, 200
  Longnose, 201
  Starry, 200

Snailfish, Lobefin, 139
  Showy, 141
  Slimy, 140
  Slipskin, 138
  Spotted, 138
  Tidepool, 138
Sole, C-O, 123
  Curlfin, 125
  Dover, 126
  English, 125
  Rex, 126
  Rock, 125
  Sand, 127
  Slender, 127
Stingray, Diamond, 203
  Pelagic, 203
  Round, 203
Sunfish, Ocean, 149
Surfperch, Barred, 180
  Redtail, 180
  Shiner, 177
  Walleye, 181

T

Thornback, 201
Thornyhead, Shortspine, 55
Tomcod, Pacific, 110
Topsmelt, 165
Treefish, 47
Triggerfish, Finescale, 159
  Redtail, 157
Tube-snout, 149
Tuna, Bigeye, 170
  Bluefin, 170
  Longfin, 170
  Yellowfin, 171
Turbot, Hornyhead, 123

W

Warbonnet, Decorated, 87
  Mosshead, 85
Whitefish, Ocean, 113
Wolf-eel, 81
Wrasse, Rock, 153
Wrymouth, Giant, 81

Y

Yellowtail, 167

SCIENTIFIC NAME INDEX

A

Agonopsis vulsa, 135
Albula vulpes, 177
Alloclinus holderi, 103
Amphistichus argenteus, 180
    rhodoterus, 180
Anarrhichthys ocellatus, 81
Anisotremus davidsoni, 185
Anoplagonus inermis, 137
Anoplarchus insignis, 85
    purpurescens, 85
Antennarius avalonis, 139
Apodichthys flavidus, 89
    fucorum, 89
Apogon guadalupensis, 149
Apristurus brunneus, 197
Artedius corallinus, 63
    fenestralis, 61
    harringtoni, 63
    lateralis, 60
Atherinops affinis, 165
    californiensis, 164
Aulorhynchus flavidus, 149

B

Balistes polylepis, 159
Blepsias cirrhosus, 75
Brachyistius frenatus, 179
Brosmophycis marginata, 113
Bryx dunckeri, 144

C

Caranx caballus, 167
Carcharodon carcharias, 191
Caulolatilus princeps, 113
Cebidichthys violaceus, 83
Cephaloscyllium ventriosum, 195
Cetorhinus maximus, 197
Chaenopsis alepidota, 99
Chaetodon falcifer, 153
Cheilotrema saturnum, 187
Chilara taylori, 79
Chilomycterus affinis, 159
Chirolophis decoratus, 87
    nugator, 85
Chitonotus pugetensis, 67
Chromis punctipinnis, 155
Citharichthys sordidus, 131
    stigmaeus, 131
    xanthostigma, 130
Clinocottus analis, 71
    globiceps, 70
    recalvus, 70
Coryphaena hippurus, 173
Coryphopterus nicholsi, 117
Cryptacanthodes giganteus, 81
Cryptotrema corallinum, 103
Cymatogaster aggregata, 177
Cypselurus californicus, 175
    hubbsi, 174

D

Dasyatis dipterura, 203
    violacea, 203
Dasycottus setiger, 59
Decapterus scombrinus, 168
Diodon holocanthus, 159
    hystrix, 161

E

Elops affinis, 175
Embiotoca jacksoni, 181
    lateralis, 183
Engraulis mordax, 165
Enophrys bison, 59
Eopsetta exilis, 127
Eptatretus stouti, 93
Errex zachirus, 126
Euleptorhamphus viridis, 174
Eumicrotremus orbis, 139

G

Gadus macrocephalus, 111
Genyonemus lineatus, 186
Gibbonsia elegans, 105
    metzi, 100
    montereyensis, 101
Girella nigricans, 187
Gnathophis catalinensis, 79
Gobiesox maeandricus, 141
Gymnothorax mordax, 79

H

Halichoeres semicinctus, 153
Hemilepidotus hemilepidotus, 61
    spinosus, 60
Hemiramphus saltator, 175
Heterodontus francisci, 195
Heterostichus rostratus, 101
Hexagrammos decagrammus, 107
    lagocephalus, 107
    stelleri, 109
Hexanchus griseus, 193
Hippocampus ingens, 144
Hippoglossus stenolepis, 129
Hydrolagus colliei, 157
Hyperprosopon argenteum, 181
Hyporhamphus rosae, 174
    unifasciatus, 174
Hypsoblennius gentilis, 114
    gilberti, 115
    jenkinsi, 114
Hypsurus caryi, 181
Hypsypops rubicundus, 155

I

Icelinus borealis, 67
    filamentosus, 69
    tenuis, 69
Isurus oxyrinchus, 191

J

Jordania zonope, 71

K

Kyphosus analogus, 185

L

Leiocottus hirundo, 73
Lepidogobius lepidus, 119
Leptocottus armatus, 57
Liparis callyodon, 138
    florae, 138
    fucensis, 138
    greeni, 139
    mucosus, 140
    pulchellus, 141
Lumpenus sagitta, 87
Lycodes brevipes, 88
    diapterus, 88
Lycodopsis pacifica, 89
Lythrypnus dalli, 119
    zebra, 119

M

Makaira nigricans, 170
Manta birostris, 205
Medialuna californiensis, 185
Merluccius productus, 110

Microgadus proximus, 110
Microstomus pacificus, 126
Mola mola, 149
Mulloidichthys dentatus, 177
Mycteroperca jordani, 25
    xenarcha, 23
Myliobatis californica, 205
Myoxocephalus polyacanthocephalus, 55

N

Naucrates ductor, 169
Nautichthys oculofasciatus, 75
Neoclinus blanchardi, 97
    stephensae, 99
    uninotatus, 97
Notorynchus cepedianus, 192

O

Odontopyxis trispinosa, 137
Oligopus diagrammus, 115
Ophiodon elongatus, 109
Orthonopias triacis, 65
Oxyjulis californica, 151
Oxylebius pictus, 105

P

Paralabrax clathratus, 25
    maculatofasciatus, 25
    nebulifer, 27
Paralichthys californicus, 129
Phanerodon furcatus, 179
Pholis clemensi, 91
    laeta, 91
    ornata, 93
Phytichthyts chirus, 82
Platichthys stellatus, 127
Platyrhinoidis triseriata, 201
Pleuronectes bilineatus, 125
    vetulus, 125
Pleuronichthys coenosus, 123
    decurrens, 125
    verticalis, 123
Podothecus acipenserinus, 135
Porichthys myriaster, 142
    notatus, 143
Prionace glauca, 191
Pristigenys serrula, 137
Psettichthys melanostictus, 127
Psychrolutes sigalutes, 73
Ptilichthys goodei, 93

R

Raja binoculata, 201
    inornata, 200
    rhina, 201
    stellulata, 200
Rathbunella hypoplecta, 117
Rhacochilus toxotes, 183
    vacca, 183
Rhamphocottus richardsoni, 75
Rhinobatos productus, 199
Rimicola dimorpha, 140
    eigenmanni, 140
    muscarum, 141
Roncador stearnsii, 186
Ronquilus jordani, 117

S

Sardinops sagax, 165
Scomber japonicus, 169
Scomberomorus sierra, 171
Scorpaena guttata, 53
    mystes, 53
Scorpaenichthys marmoratus, 57
Scorpaenodes xyris, 55
Sebastes atrovirens, 37
    auriculatus, 37
    brevispinis, 41
    carnatus, 29
    caurinus, 31
    chrysomelas, 27
    ciliatus, 43
    constellatus, 45
    dalli, 29
    elongatus, 49
    emphaeus, 35
    entomelas, 39
    flavidus, 39
    hopkinsi, 35
    maliger, 29
    melanops, 43
    miniatus, 33
    mystinus, 43
    nebulosus, 27
    nigrocinctus, 47
    paucispinis, 41
    pinniger, 35
    proriger, 33
    rastrelliger, 37
    rosaceus, 45
    ruberrimus, 31
    rubrivinctus, 49
    serranoides, 39
    serriceps, 47
    umbrosus, 45
Sebastolobus alascanus, 55
Semicossyphus pulcher, 151
Seriola lalandi, 167
    rivoliana, 167
Sphoeroides annulatus, 161
Sphyraena argentea, 173
Sphyrna lewini, 197
    mokarran, 196
    zygaena, 196
Squalus acanthias, 193
Squatina californica, 199
Stereolepis gigas, 23
Synchirus gilli, 73
Syngnathus auliscus, 142
    californiensis, 142
    euchrous, 145
    exilis, 142
    leptorhynchus, 143
Synodus lucioceps, 113

T

Tetrapturus audax, 171
Theragra chalcogramma, 111
Thunnus alalunga, 170
    albacares, 171
    obesus, 170
    thynnus, 170
Torpedo californica, 205
Trachurus symmetricus, 169
Triakis semifasciata, 193
Triglops macellus, 65

U

Ulvicola sanctaerosae, 88
Umbrina roncador, 186
Urolophus halleri, 203

X

Xanthichthys mento, 157
Xeneretmus latifrons, 135
Xiphister atropurpureus, 83
    mucosus, 82
Zaniolepis latipinnis, 111
Zapteryx exasperata, 199

PERSONAL RECORD OF FISH SIGHTINGS

1. HEAVY BODY/LARGE LIPS
Seabass - Rockfish

No.	Name	Page	Date	Location	Notes
	GIANT SEA BASS Stereolepis gigas	23			
	BROOMTAIL GROUPER Mycteroperca xenarcha	23			
	GULF GROUPER Mycteroperca jordani	25			
	KELP BASS Paralabrax clathratus	25			
	SPOTTED SAND BASS Paralabrax maculatofasciatus	25			
	BARRED SAND BASS Paralabrax nebulifer	27			
	CHINA ROCKFISH Sebastes nebulosus	27			
	BLACK-AND-YELLOW ROCKFISH Sebastes chrysomelas	27			
	GOPHER ROCKFISH Sebastes carnatus	29			
	QUILLBACK ROCKFISH Sebastes maliger	29			
	CALICO ROCKFISH Sebastes dalli	29			
	COPPER ROCKFISH Sebastes caurinus	31			
	YELLOWEYE ROCKFISH Sebastes ruberrimus	31			
	REDSTRIPE ROCKFISH Sebastes proriger	33			
	VERMILION ROCKFISH Sebastes miniatus	33			
	CANARY ROCKFISH Sebastes pinniger	35			
	PUGET SOUND ROCKFISH Sebastes emphaeus	35			
	SQUARESPOT ROCKFISH Sebastes hopkinsi	35			
	BROWN ROCKFISH Sebastes auriculatus	37			
	GRASS ROCKFISH Sebastes rastrelliger	37			
	KELP ROCKFISH Sebastes atrovirens	37			
	YELLOWTAIL ROCKFISH Sebastes flavidus	39			
	OLIVE ROCKFISH Sebastes serranoides	39			
	WIDOW ROCKFISH Sebastes entomelas	39			
	SILVERGRAY ROCKFISH Sebastes brevispinis	41			
	BOCACCIO Sebastes paucispinis	41			
	DUSKY ROCKFISH Sebastes ciliatus	43			
	BLACK ROCKFISH Sebastes melanops	43			

No.	Name	Page	Date	Location	Notes
	BLUE ROCKFISH Sebastes mystinus	43			
	HONEYCOMB ROCKFISH Sebastes umbrosus	45			
	ROSY ROCKFISH Sebastes rosaceus	45			
	STARRY ROCKFISH Sebastes constellatus	45			
	TIGER ROCKFISH Sebastes nigrocinctus	47			
	TREEFISH Sebastes serriceps	47			
	FLAG ROCKFISH Sebastes rubrivinctus	49			
	GREENSTRIPED ROCKFISH Sebastes elongatus	49			

2. BULBOUS, SPINY-HEADED BOTTOM-DWELLERS
Scorpionfish - Sculpin

No.	Name	Page	Date	Location	Notes
	CALIFORNIA SCORPIONFISH Scorpaena guttata	53			
	STONE SCORPIONFISH Scorpaena mystes	53			
	RAINBOW SCORPIONFISH Scorpaenodes xyris	55			
	SHORTSPINE THORNYHEAD Sebastolobus alascanus	55			
	GREAT SCULPIN Myoxocephalus polyacanthocephalus	55			
	CABEZON Scorpaenichthys marmoratus	57			
	PACIFIC STAGHORN SCULPIN Leptocottus armatus	57			
	BUFFALO SCULPIN Enophrys bison	59			
	SPINYHEAD SCULPIN Dasycottus setiger	59			
	RED IRISH LORD Hemilepidotus hemilepidotus	61			
	BROWN IRISH LORD Hemilepidotus spinosus	60			
	PADDED SCULPIN Artedius fenestralis	61			
	SMOOTHHEAD SCULPIN Artedius lateralis	60			
	SCALYHEAD SCULPIN Artedius harringtoni	63			
	CORALLINE SCULPIN Artedius corallinus	63			
	SNUBNOSE SCULPIN Orthonopias triacis	65			
	ROUGHSPINE SCULPIN Triglops macellus	65			
	ROUGHBACK SCULPIN Chitonotus pugetensis	67			
	NORTHERN SCULPIN Icelinus borealis	67			
	SPOTFIN SCULPIN Icelinus tenuis	69			
	THREADFIN SCULPIN Icelinus filamentosus	69			

No.	Name	Page	Date	Location	Notes
	LONGFIN SCULPIN Jordania zonope	71			
	WOOLLY SCULPIN Clinocottus analis	71			
	MOSSHEAD SCULPIN Clinocottus globiceps	70			
	BALD SCULPIN Clinocottus recalvus	70			
	SOFT SCULPIN Psychrolutes sigalutes	73			
	MANACLED SCULPIN Synchirus gilli	73			
	LAVENDER SCULPIN Leiocottus hirundo	73			
	SAILFIN SCULPIN Nautichthys oculofasciatus	75			
	SILVERSPOTTED SCULPIN Blepsias cirrhosus	75			
	GRUNT SCULPIN Rhamphocottus richardsoni	75			

3. EELS AND EEL-LIKE BOTTOM-DWELLERS
Prickleback - Gunnel - Others

No.	Name	Page	Date	Location	Notes
	CALIFORNIA MORAY Gymnothorax mordax	79			
	CATALINA CONGER Gnathophis catalinensis	79			
	SPOTTED CUSK-EEL Chilara taylori	79			
	WOLF-EEL Anarrhichthys ocellatus	81			
	GIANT WRYMOUTH Cryptacanthodes giganteus	81			
	MONKEYFACE PRICKLEBACK Cebidichthys violaceus	83			
	BLACK PRICKLEBACK Xiphister atropurpureus	83			
	ROCK PRICKLEBACK Xiphister mucosus	82			
	RIBBON PRICKLEBACK Phytichthyts chirus	82			
	HIGH COCKSCOMB Anoplarchus purpurescens	85			
	SLENDER COCKSCOMB Anoplarchus insignis	85			
	MOSSHEAD WARBONNET Chirolophis nugator	85			
	DECORATED WARBONNET Chirolophis decoratus	87			
	SNAKE PRICKLEBACK Lumpenus sagitta	87			
	BLACKBELLY EELPOUT Lycodopsis pacifica	89			
	BLACK EELPOUT Lycodes diapterus	88			
	SHORTFIN EELPOUT Lycodes brevipes	88			
	ROCKWEED GUNNEL Apodichthys fucorum	89			
	KELP GUNNEL Ulvicola sanctaerosae	88			

No.	Name	Page	Date	Location	Notes
	PENPOINT GUNNEL Apodichthys flavidus	89			
	LONGFIN GUNNEL Pholis clemensi	91			
	CRESCENT GUNNEL Pholis laeta	91			
	SADDLEBACK GUNNEL Pholis ornata	93			
	PACIFIC HAGFISH Eptatretus stouti	93			
	QUILLFISH Ptilichthys goodei	93			

4. ELONGATED BOTTOM-DWELLERS
Fringehead - Kelpfish - Greenling - Others

No.	Name	Page	Date	Location	Notes
	SARCASTIC FRINGEHEAD Neoclinus blanchardi	97			
	ONESPOT FRINGEHEAD Neoclinus uninotatus	97			
	YELLOWFIN FRINGEHEAD Neoclinus stephensae	99			
	ORANGETHROAT PIKEBLENNY Chaenopsis alepidota	99			
	GIANT KELPFISH Heterostichus rostratus	101			
	CREVICE KELPFISH Gibbonsia montereyensis	101			
	STRIPED KELPFISH Gibbonsia metzi	100			
	DEEPWATER BLENNY Cryptotrema corallinum	103			
	ISLAND KELPFISH Alloclinus holderi	103			
	SPOTTED KELPFISH Gibbonsia elegans	105			
	PAINTED GREENLING Oxylebius pictus	105			
	ROCK GREENLING Hexagrammos lagocephalus	107			
	KELP GREENLING Hexagrammos decagrammus	107			
	WHITESPOTTED GREENLING Hexagrammos stelleri	109			
	LINGCOD Ophiodon elongatus	109			
	LONGSPINE COMBFISH Zaniolepis latipinnis	111			
	PACIFIC COD Gadus macrocephalus	111			
	PACIFIC TOMCOD Microgadus proximus	110			
	WALLEYE POLLOCK Theragra chalcogramma	111			
	PACIFIC HAKE Merluccius productus	110			
	CALIFORNIA LIZARDFISH Synodus lucioceps	113			
	OCEAN WHITEFISH Caulolatilus princeps	113			
	RED BROTULA Brosmophycis marginata	113			
	PURPLE BROTULA Oligopus diagrammus	115			

No.	Name	Page	Date	Location	Notes
	ROCKPOOL BLENNY Hypsoblennius gilberti	115			
	BAY BLENNY Hypsoblennius gentilis	114			
	MUSSEL BLENNY Hypsoblennius jenkinsi	114			
	STRIPEDFIN RONQUIL Rathbunella hypoplecta	117			
	NORTHERN RONQUIL Ronquilus jordani	117			
	BLACKEYE GOBY Coryphopterus nicholsi	117			
	BAY GOBY Lepidogobius lepidus	119			
	BLUEBANDED GOBY Lythrypnus dalli	119			
	ZEBRA GOBY Lythrypnus zebra	119			

5. FLATFISH/BOTTOM-DWELLERS
Flounder - Turbot - Sole - Halibut - Sanddab

	Name	Page	Date	Location	Notes
	HORNYHEAD TURBOT Pleuronichthys verticalis	123			
	C-O SOLE Pleuronichthys coenosus	123			
	CURLFIN SOLE Pleuronichthys decurrens	125			
	ROCK SOLE Pleuronectes bilineatus	125			
	ENGLISH SOLE Pleuronectes vetulus	125			
	SAND SOLE Psettichthys melanostictus	127			
	SLENDER SOLE Eopsetta exilis	127			
	REX SOLE Errex zachirus	126			
	DOVER SOLE Microstomus pacificus	126			
	STARRY FLOUNDER Platichthys stellatus	127			
	PACIFIC HALIBUT Hippoglossus stenolepis	129			
	CALIFORNIA HALIBUT Paralichthys californicus	129			
	PACIFIC SANDDAB Citharichthys sordidus	131			
	LONGFIN SANDDAB Citharichthys xanthostigma	130			
	SPECKLED SANDDAB Citharichthys stigmaeus	131			

6. ODD-SHAPED BOTTOM-DWELLERS
Poacher - Snailfish - Pipefish & Seahorse - Others

	Name	Page	Date	Location	Notes
	N. SPEARNOSE POACHER Agonopsis vulsa	135			
	STURGEON POACHER Podothecus acipenserinus	135			
	BLACKTIP POACHER Xeneretmus latifrons	135			

No.	Name	Page	Date	Location	Notes
	PYGMY POACHER Odontopyxis trispinosa	137			
	SMOOTH ALLIGATORFISH Anoplagonus inermis	137			
	POPEYE CATALUFA Pristigenys serrula	137			
	ROUGHJAW FROGFISH Antennarius avalonis	139			
	PACIFIC SPINY LUMPSUCKER Eumicrotremus orbis	139			
	LOBEFIN SNAILFISH Liparis greeni	139			
	TIDEPOOL SNAILFISH Liparis florae	138			
	SPOTTED SNAILFISH Liparis callyodon	138			
	SLIPSKIN SNAILFISH Liparis fucensis	138			
	SHOWY SNAILFISH Liparis pulchellus	141			
	SLIMY SNAILFISH Liparis mucosus	140			
	NORTHERN CLINGFISH Gobiesox maeandricus	141			
	KELP CLINGFISH Rimicola muscarum	141			
	SOUTHERN CLINGFISH Rimicola dimorpha	140			
	SLENDER CLINGFISH Rimicola eigenmanni	140			
	PLAINFIN MIDSHIPMAN Porichthys notatus	143			
	SPECKLEFIN MIDSHIPMAN Porichthys myriaster	142			
	BAY PIPEFISH Syngnathus leptorhynchus	143			
	BARCHEEK PIPEFISH Syngnathus exilis	142			
	BARRED PIPEFISH Syngnathus auliscus	142			
	KELP PIPEFISH Syngnathus californiensis	142			
	CHOCOLATE PIPEFISH Syngnathus euchrous	145			
	PUGNOSE PIPEFISH Bryx dunckeri	144			
	PACIFIC SEAHORSE Hippocampus ingens	144			

7. ODD-SHAPED & OTHER SWIMMERS
Wrasse - Others

No.	Name	Page	Date	Location	Notes
	OCEAN SUNFISH Mola mola	149			
	TUBE-SNOUT Aulorhynchus flavidus	149			
	GUADALUPE CARDINALFISH Apogon guadalupensis	149			
	CALIFORNIA SHEEPHEAD Semicossyphus pulcher	151			
	SENORITA Oxyjulis californica	151			

No.	Name	Page	Date	Location	Notes
	ROCK WRASSE Halichoeres semicinctus	153			
	SCYTHE BUTTERFLYFISH Chaetodon falcifer	153			
	BLACKSMITH Chromis punctipinnis	155			
	GARIBALDI Hypsypops rubicundus	155			
	SPOTTED RATFISH Hydrolagus colliei	157			
	REDTAIL TRIGGERFISH Xanthichthys mento	157			
	FINESCALE TRIGGERFISH Balistes polylepis	159			
	PACIFIC BURRFISH Chilomycterus affinis	159			
	BALLOONFISH Diodon holocanthus	159			
	PORCUPINEFISH Diodon hystrix	161			
	BULLSEYE PUFFER Sphoeroides annulatus	161			

8. SILVERY SWIMMERS
Jack - Mackerel - Surfperch - Others

No.	Name	Page	Date	Location	Notes
	PACIFIC SARDINE Sardinops sagax	165			
	NORTHERN ANCHOVY Engraulis mordax	165			
	TOPSMELT Atherinops affinis	165			
	JACKSMELT Atherinops californiensis	164			
	GREEN JACK Caranx caballus	167			
	YELLOWTAIL Seriola lalandi	167			
	ALMACO JACK Seriola rivoliana	167			
	PILOTFISH Naucrates ductor	169			
	JACK MACKEREL Trachurus symmetricus	169			
	MEXICAN SCAD Decapterus scombrinus	168			
	CHUB MACKEREL Scomber japonicus	169			
	PACIFIC SIERRA Scomberomorus sierra	171			
	YELLOWFIN TUNA Thunnus albacares	171			
	LONGFIN TUNA Thunnus alalunga	170			
	BIGEYE TUNA Thunnus obesus	170			
	BLUEFIN TUNA Thunnus thynnus	170			
	STRIPED MARLIN Tetrapturus audax	171			
	BLUE MARLIN Makaira nigricans	170			

No.	Name	Page	Date	Location	Notes
	DOLPHIN Coryphaena hippurus	173			
	PACIFIC BARRACUDA Sphyraena argentea	173			
	MACHETE Elops affinis	175			
	CALIFORNIA FLYINGFISH Cypselurus californicus	175			
	BLOTCHWING FLYINGFISH Cypselurus hubbsi	174			
	LONGFIN HALFBEAK Hemiramphus saltator	175			
	CALIFORNIA HALFBEAK Hyporhamphus rosae	174			
	SILVERSTRIPE HALFBEAK Hyporhamphus unifasciatus	174			
	RIBBON HALFBEAK Euleptorhamphus viridis	174			
	BONEFISH Albula vulpes	177			
	MEXICAN GOATFISH Mulloidichthys dentatus	177			
	SHINER PERCH Cymatogaster aggregata	177			
	WHITE SEAPERCH Phanerodon furcatus	179			
	KELP PERCH Brachyistius frenatus	179			
	WALLEYE SURFPERCH Hyperprosopon argenteum	181			
	BLACK PERCH Embiotoca jacksoni	181			
	REDTAIL SURFPERCH Amphistichus rhodoterus	180			
	BARRED SURFPERCH Amphistichus argenteus	180			
	RAINBOW SEAPERCH Hypsurus caryi	181			
	STRIPED SEAPERCH Embiotoca lateralis	183			
	PILE PERCH Rhacochilus vacca	183			
	RUBBERLIP SEAPERCH Rhacochilus toxotes	183			
	SARGO Anisotremus davidsoni	185			
	BLUE-BRONZE CHUB Kyphosus analogus	185			
	HALFMOON Medialuna californiensis	185			
	OPALEYE Girella nigricans	187			
	BLACK CROAKER Cheilotrema saturnum	187			
	YELLOWFIN CROAKER Umbrina roncador	186			
	SPOTFIN CROAKER Roncador stearnsii	186			
	WHITE CROAKER Genyonemus lineatus	186			

9. SHARKS & RAYS

No.	Name	Page	Date	Location	Notes
	BLUE SHARK Prionace glauca	191			
	WHITE SHARK Carcharodon carcharias	191			
	SHORTFIN MAKO Isurus oxyrinchus	191			
	SPINY DOGFISH Squalus acanthias	193			
	LEOPARD SHARK Triakis semifasciata	193			
	SIXGILL SHARK Hexanchus griseus	193			
	SEVENGILL SHARK Notorynchus cepedianus	192			
	HORN SHARK Heterodontus francisci	195			
	SWELL SHARK Cephaloscyllium ventriosum	195			
	BROWN CAT SHARK Apristurus brunneus	197			
	BASKING SHARK Cetorhinus maximus	197			
	SCALLOPED HAMMERHEAD Sphyrna lewini	197			
	GREAT HAMMERHEAD Sphyrna mokarran	196			
	SMOOTH HAMMERHEAD Sphyrna zygaena	196			
	ANGEL SHARK Squatina californica	199			
	SHOVELNOSE GUITARFISH Rhinobatos productus	199			
	BANDED GUITARFISH Zapteryx exasperata	199			
	BIG SKATE Raja binoculata	201			
	STARRY SKATE Raja stellulata	200			
	LONGNOSE SKATE Raja rhina	201			
	CALIFORNIA SKATE Raja inornata	200			
	THORNBACK Platyrhinoidis triseriata	201			
	DIAMOND STINGRAY Dasyatis dipterura	203			
	PELAGIC STINGRAY Dasyatis violacea	203			
	ROUND STINGRAY Urolophus halleri	203			
	PACIFIC ELECTRIC RAY Torpedo californica	205			
	BAT RAY Myliobatis californica	205			
	MANTA Manta birostris	205			

THE REEF SET
by
Paul Humann

REEF FISH IDENTIFICATION FLORIDA-CARIBBEAN-BAHAMAS 2nd Edition
The book that revolutionized fishwatching just got better!
424 pp, 670 color plates. **$39.95**

REEF CREATURE IDENTIFICATION FLORIDA-CARIBBEAN-BAHAMAS
More than 30 marine life scientists from eight nations collaborated with the author to compile the most comprehensive and accurate visual identification guide of reef invertebrates ever published.
344 pp, 478 color plates. **$37.95**

REEF CORAL IDENTIFICATION FLORIDA-CARIBBEAN-BAHAMAS
Stony, soft, fire and black corals. Includes an appendix of marine plants.
252 pp, 475 color plates. **$32.95**

Shelf Case for the Three Volume REEF SET **$10.00**
Three Volume REEF SET with Shelf Case **$115.00**
Weather-resistant, canvas Traveler's Case for the REEF SET **$25.00**
Three Volume REEF SET with Traveler's Case **$130.00**

REEF FISH IDENTIFICATION GALAPAGOS
Finally, a comprehensive fish identification guide for the Galapagos Islands — the world's most spectacular natural aquarium. 260 beautiful color plates display the famous archipelago's fabulous fish life in an easy-to-use, quick reference format.
Comb binding, 200 pp, 6"X 9". **$32.95**

DIVING GUIDE TO UNDERWATER FLORIDA 9th Edition
The most complete and up-to-date information available about diving in Florida, with 50 maps and directions to over 500 superb diving locations.
Soft cover, 340 pp, 6"X 9". **$16.95**

To order any of the above products call 1-800-737-6558, or write to:

New World Publications
1861 Cornell Rd. Dept. C, Jacksonville, Florida 32207

Please include $5.00 for shipping and handling for shipments within the U.S.
VISA and Master Card accepted.